Economy and Ecology:
Towards Sustainable Development

Economy and Ecology: Towards Sustainable Development

edited by

F. Archibugi

Department of Planning Theory,
University of Naples,
Naples, Italy

and

P. Nijkamp

Faculty of Economics,
Free University,
Amsterdam, The Netherlands

Kluwer Academic Publishers
Dordrecht / Boston / London

ISBN 0-7923-0477-2

Published by Kluwer Academic Publishers,
P.O. Box 17, 3300 AA Dordrecht, The Netherlands.

Kluwer Academic Publishers incorporates
the publishing programmes of
D. Reidel, Martinus Nijhoff, Dr W. Junk and MTP Press.

Sold and distributed in the U.S.A. and Canada
by Kluwer Academic Publishers,
101 Philip Drive, Norwell, MA 02061, U.S.A.

In all other countries, sold and distributed
by Kluwer Academic Publishers Group,
P.O. Box 322, 3300 AH Dordrecht, The Netherlands.

Second printing 1990

printed on acid free paper

Printed in The Netherlands

TABLE OF CONTENTS

Preface

After a period of relative silence, recent years have been marked by an upswing of interest in environmental issues. The publication of the report of the World Commission on Environment and Development on 'Our Common Future' (1987) has acted as a catalyst for a revival of the environmental awareness, not only regarding local and daily pollution problems, but also -and in particular- regarding global environmental decay and threats to a sustainable development.

In a recent study by W.M. Stigliani et al., on 'Future Environments for Europe' (Executive Report 15, IIASA, Laxenburg, 1989) the environmental implications of various alternative socioeconomic development pathways with respect to eleven environmental issues that could become major problems in the future are analysed. These issues include:

- Managing water resources in an era of climate change.
- Acidification of soils and lakes in Europe.
- Long-term forestry management and the possibility of a future shortfall in wood supply.
- Areas of Europe marginalized by mainstream economic and agricultural development.
- Sea level rise.
- Chemical pollution of coastal waters.
- Toxic materials buildup and the potential for chemical time bombs.
- Non-point-source emissions of potentially toxic substances.
- Transportation growth versus air quality.
- Decreasing multi-functionally of land owing to urban and suburban land development.
- Increasing summer demand for electricity, and the impact on air quality.

Compared to the environmental issues that have usually been covered in the environmental economics literature, the latter set of issues is much more global, long-term and far-reaching in nature. Nevertheless, various important trade-offs have still remained the same, and therefore it is important to explore whether a resource management based on ecologically sustainable economic development can be achieved through the application of sound economic analysis, exemplified, inter alia, by social cost-benefit analysis, multi-criteria analysis and integrated economic-ecological simulation analysis. Special attention has to be given here to the dynamics of resource systems management.

Consequently, economic efficiency, ecological sustainability and multi-temporal trade-off would have to be integrated in one evaluation system. Now it is interesting to observe that in various current discussions on resource use the rich harvest from previous analyses in economics and ecology is sometimes neglected. In order to explore the potential of the economic-ecological analytical heritage from the past years, an international conference was held in Milan in spring 1988. Various contributions in the present volume stem from this meeting. The aim was to show that economic and ecological guidelines may be meaningful vehicles for achieving a balanced and sustainable development of our planet.

Our thanks and appreciation have to be expressed to the Italian Minister for the Environment, Giorgio Ruffolo, who took the initiative of promoting this environmental conference in Milan, in March 1988, on the theme of the relationship between environment and development.

This meeting attracted world-wide attention and was attended by numerous politicians and scientists from all over the world. The editors of this volume have decided to select from all written and oral presentations a set of papers which provide a refreshing look at the interface of economics and ecology. It is our hope that this panoramic overview will induce further steps on the way toward sustainable development.

The editors wish to thank Rita Hittema for her painstaking key-boarding work on this manuscript and Winny Browne for her careful editing of the first version of the present book. We also wish to express our appreciation to all contributors of this book who shared our task to produce a 'sustainable' publication.

June, 1989

Rome — Franco Archibugi

Amsterdam — Peter Nijkamp

LIST OF CONTRIBUTORS

F. Archibugi
Centro di Studi e Piani Economici
Via Federico Cassitto, 110
00134 Roma
Italy

R.U. Ayres
Carnegie-Mellon University
Pittsburgh, Pa
USA

M. Bresso
Dept. of Political Economics
University of Turin
Via S. Ottavio, 20
Turin
Italy

B. Commoner
Center for the Biology of Natural Systems
Queens College
State University of New York
Flushing NY
USA

H.E. Daly
World Bank
1818 H Street
Washington, DC
USA

P. Dasgupta
Faculty of Economics
Siegewick Avenue
Cambridge University
Cambridge CB3 9D0
England

K.A. Fox
Dept. of Economics
Iowa State University
Ames, Iowa 50011
USA

D.E. James
School of Economic and Financial Studies
Macquarie University
North Ryde NSW
2109 Australia

A.V. Kneese
Resources for the Future
1616 P Street NW
Washington, D.C. 20036
USA

H. Leeflang
State Physical Planning Agency
Willem Witsenplein 6
The Hague
The Netherlands

S. Lombardini
University of Torino
c/o Via Roaschia, 137
I-10023 Chieri, Torino
Italy

J. MacNeill
The Institute for Research on Public Policy
275 Slater Street
Ottawa Ont KIP 5H9
Canada

C.-G. Mäler
Stockholm School of Economics
P.O. Box 6501
S-11383 Stockholm
Sweden

R.E. Munn
Environment Division IIASA
A-2361 Laxenburg
Austria

P. Nijkamp
Dept. of Economics
Free University
P.O. Box 7161
1007 MC Amsterdam
The Netherlands

J.B. Opschoor
Institute for Environmental Studies
Free University
P.O. Box 7161
1007 MC Amsterdam
The Netherlands

G. Scimemi
Environment Directorate
OECD
2, Rue André-Pascal
75775 Paris Cedex 16
France

U. Simonis
Science Center Berlin
Reichpietshufer 50
D-1000 Berlin
Federal Republic of Germany

F.J. Soeteman
Economic and Social Institute
Free University
P.O. Box 7161
1007 MC Amsterdam
The Netherlands

L.E. Susskind
32 Jericho Hill Road
Southbourough, MA 017772
USA

K. Uno
Institute of Socio-Economic Planning
University of Tsukuba
Sakura, Ibaraki 305
Japan

Chapter 1

THE CHALLENGE OF SUSTAINABLE DEVELOPMENT

F. Archibugi

P. Nijkamp

F.J. Soeteman

Economy and Ecology: A Helicopter's View

Environmental decay is not exclusively a phenomenon of our century. The Greek philosopher Plato already complains in his Critias about the landscape changes in Attica which had transformed the environment into "... bones of wasted body ... richer and softer parts of the soil having fallen away, and the mere skeleton being left" (cited in Clark, 1986, p. 6). Also in many other countries one observes many examples of earlier soil erosion which -as a result of agricultural and forestry activities- has affected the landscape in all time periods between nomadic cultures and modern high-tech agriculture (Wilkinson, 1973).

Even so, until the beginning of the twentieth century, in general, only relatively modest environmental changes were taking place, as the prevailing technological and economic system was unable to alter environmental conditions on earth dramatically. However, especially after World War II mankind's capacity to destroy our habitat has increased significantly, partly as a result of radical technological changes (generating huge amounts of air, water and soil pollutants including many toxic materials), partly as a result of the rise in world population (and its subsequent rise in consumption and mobility patterns). The strive for a concerted development of the economy and ecology -based on a coherent and integrated viewpoint- has stimulated many social scientists to adopt systemic notions and concepts for achieving a balance between natural and socio-economic systems. The functioning of such natural and socio-economic systems has in their view to be studied from the angle of materials inputs and outputs of all production and consumption processes. In this context, new sub-disciplines such as human ecology and environmental economics have come to the fore, in which serious attempts have been made to ensure a merger between economics and ecology (see also Nijkamp, 1978).

An important contribution to the integration of economics and ecology began simply with a reflection on the principle of the materials balance for resources (extracted or collected, transformed, consumed and emitted) and on the need to take account of an economic viewpoint of such processes (cf. Georgescu-Roegen, 1971 or Kneese et al., 1970).

On the other hand, much thought has been given to the impact that an integration of economics and ecology could (or should) have on economic thinking or on theory building itself (cf. Boulding, 1966; Daly, 1973; Georgescu-Roegen, 1971, 1973; and Kapp, 1970).

Next, attempts have been made to build economic and social accounting systems which could incorporate the measurement of economic welfare and performance together with the measurement of environmental indicators and performance (Fox, 1971, 1985; Juster et al., 1979; Net National Welfare Commission of Japan, 1973; and Nordhaus and Tobin, 1972).

Besides, analytical schemes have been experimented to include, in the general ,systems of models for planning purposes at national and regional levels, new models that were also able to include environmental evaluation in the decision process of planning (Archibugi, 1974 and Leontief, 1970, 1973).

It is also noteworthy that various theories have been proposed regarding conventional schemes of programme and project evaluation such as cost-benefit analysis, methods and techniques of environmental impact evaluation, and the evaluation of social impacts (Dasgupta and Pearce, 1972 and Pearce, 1978).

Ways and means of 'internalising' environmental damage in production costs (and, likewise, in the production function) have also been discussed, as well as the cost of its management and elimination, with the main emphasis on different systems of taxation (Barde and Gerelli, 1977; Baumol and Oates, 1975; and OECD, 1975).

The integration of economics with ecology has also been approached from the viewpoint of land-use, where economic and ecological processes have the most disruptive effects, and in urban environments (Doxiadis, 1968; and Mesarovic and Reisman, 1972).

The interaction between economics and ecology has next been dealt with for situations with global risks and uncertainties (for instance, Bremer et al., 1987; Clark and Munn, 1986; Meadows et al., 1972; and Mesarovic and Pestel, 1974).

All these research directions are to some extent oriented toward an integrative approach to economics and ecology, i.e. to the design of methods, techniques and tools for a cohesive evaluation, by which values and objectives in both the environmental and the socio-economic area are simultaneously taken into consideration.

Meanwhile, an increasing political interest has developed regarding the disruption of the earth's natural resources and environmental decay. Despite many efforts, local, national and international policy bodies have been unsuccessful in ensuring a viable economic development trajectory that was compatible with environmental quality. For instance, the big UN Conference on the Human Environment (Stockholm, 1972) has only very moderately achieved the high goals which were set for our planet. The World Commission on Environment and Development, established by the General Assembly of the United Nations in 1983, was assigned the task to provide concrete recommendations for action on the interrelated issues of environment and development, seen from a strategic long-term viewpoint. The Commission Report (1987), "Our Common Future" (often named the Brundtland Report after the President of the Commission) is a remarkable document in that it offers a concrete hope for sustainable development.

Nevertheless, the Report is not utopian in nature. It spells out the major threats to our common future. These are inter alia

- massive poverty (causing amongst others deforestation and urban decay);
- population growth (causing amongst others serious food and housing problems);
- global warming and climatic changes (causing amongst others sea level rise and depletion of the ozone layer);
- destruction of environmental quality (causing amongst others deforestation, desertification and extinction of ecological species).

All such threats call for effective action and the Commission believes that - despite the potential catastrophes incorporated in our modern way of life - human resources, knowledge and capabilities are available to create a sustainable development. Sustainable development is defined here as paths of human progress which meet the needs and aspirations of the present generation without compromising the ability of future generations to meet their needs. It hence requires a fairer distribution of wealth within and among countries and groups in society. In this context, economic growth is not by definition a threat to sustainability, but even the only feasible weapon in the fight against poverty and disaster; with economic growth we can create the capacity to alleviate poverty and solve environmental threats. This requires economy and ecology be merged from a local to a global perspective. This issue of the compatibility of economy and ecology will be further discussed in the next section.

Economy and Ecology: Co-evolutionary Paths

In the light of the observations made in the previous section the question may be raised whether a co-evolutionary and sustainable development of economic conditions and environmental qualities is a feasible option (see also Norgaard, 1984). Such a co-evolutionary development would in our view imply a simultaneous (and preferably parallel) improvement of both the economic system and the environmental system. Or otherwise stated, it would imply a Pareto principle in which an improvement in the one system does not structurally affect the other. In a more recent article, Norgaard (1988) makes a plea for a development model which does not reflect a unilineal view on progress, but one which encompasses a broad understanding of human/environment interaction. Thus co-evolution takes for granted a balance between economic development (all quantitative and qualitative changes in the economy that lead to a positive contribution to welfare) and ecological sustainability (all quantitative and qualitative environmental changes that serve to improve the quality of an ecosystem and hence have also a positive influence on welfare).

In itself, the concept of 'sustainability' of development utilised in the Brundtland Report is not a special novelty in the evolution of economic thinking. The wish to take account of the needs of future generations while, at the same time, programming to satisfy the needs of present generations refers essentially to the opportunity of taking account of objectives in both the long and the medium term; this is an elementary principle of any kind of economic, social and physical planning. However, in the Brundtland Report a plea is made to remove persistent obstacles regarding inertia, routine and ineffectiveness (either governmental or non-governmental), that impedes a faster enforcement and management of an environmental policy.

It is noteworthy that we use here a broad welfare concept in the sense of all (individual or collective) utility derived from the availability or use of scarce resources, no matter whether such utility attributes can be measured in monetary terms or not (the so-called formal welfare concept; see Nijkamp and Soeteman, 1988). Consequently, also environmental measures against the use of toxic materials, ionizing radiation or traffic accidents may be regarded as arguments of a welfare function.

In a recent article, Pearce (1988) adopts a specific position in the co-evolution debate by claiming that a further decline in natural resources beyond a point that we have already reached is actually injurious to economic development. Assuming that sustainability is a necessary condition for economic growth, he argues that the stock of "natural capital" should at least be kept constant (and preferably increased) while the economy is allowed whatever social goals are deemed appropriate. Clearly, this viewpoint raises important questions concerning the measurability of environmental quality.

In any case, the conclusion can be drawn that sustainable development cannot be measured by means of a single indicator, but needs a broad spectrum of manifest indicators.

For instance, in the framework of agricultural activities, the welfare gains from agriculture should not only be measured by value added created in the agricultural sector, but should also incorporate negative externalities regarding landscape, species diversity or eco-stability (see also Dahlberg, 1986). Clearly, various changes in land use patterns on landscapes may also be due to factors outside the realm of the agricultural system itself (e.g., climatic factors).

Both the economy and the ecology may contribute to societal welfare and incorporate some (limited) degree of substitutability. However, there are convincing indications now that labour, capital and technology are substituted for environmental production factors up to a level which is clearly unsustainable for both systems in the long run (e.g., eutrophication of surface water, or stratospheric depletion of the ozone layer). On the other hand, environmental commodities and amenities (such as forests, fish, drinking water, fossil fuels, etc.) are economically used up to a level which clearly exceeds environmental growth (production) rates; this threatens environmental regeneration potentials. Consequently, both economic and environmental systems need at least a certain minimum achievement level (or threshold value) in order to survive. In this context, Ciriacy-Wantrup (1952, p. 253) has made a plea for using a minimum bequest value in strategic environmental policies, in particular the establishment of safe minimum standards of conservation. Thus the idea of a co-evolutionary development needs a careful consideration of sustainable threshold levels for both the economic and the environmental system.

Thus strategic economic and environmental policy is essentially a risk strategy which serves to minimize the potential mismatch between economic development and ecological sustainability. Strategic considerations under such circumstances may be based on the proposition that the welfare constituents of ecosystems are directly connected to their physical, chemical or biological role in the whole system. Despite the important function of environmental goods in maintaining ecosystems processes, the socio-economic value of such goods is not always unambiguous (cf. Norton,

1986). It may be approximated amongst others by using the following guiding principles regarding the use of an environmentally valuable asset:

(a) use value: the value of an environmental amenity to generate present and future benefits;

(b) risk aversion assessment: potential users are not sure that they will ever use the environmental opportunity concerned, but do not want to lose the possibility to use (or enjoy) it in the future;

(c) quasioption value: potential users have an interest in enjoying an environmental good but are willing to forego an irreversible development in order to preserve future options concerning this environmental good;

(d) moral or existence value: non-users attach a high value to the fact that the scarce environmental commodity is maintained as such;

(e) vicarious use value: non-users want to keep a certain (public) environmental good intact, because they like it when others can enjoy this asset;

(f) bequest value: non-users see it as their moral responsibility (or altruism) to protect and maintain a certain environmental asset for future generations.

Unfortunately, our current economic apparatus has not yet managed to devise an operational methodology for analyzing all the complexities involved in ecologically sustainable economic development (see Opschoor, 1987). This is also witnessed in a statement in the World Bank Annual Report (1985):

> "Degradation and destruction of environmental systems and natural resources are now assuming massive proportions in some developing countries, threatening continued, sustainable development. It is now generally recognized that economic development itself can be an important contributing factor to growing environmental problems in the absence of appropriate safeguards. A greatly improved understanding of the natural resource base and environment systems that support national economies is needed if patterns of development that are sustainable can be determined and recommended to governments."

This lack of understanding is not amazing, because even a quick glance at the history of economic thinking shows that only a few attempts have been made to position natural resources at the heart of economics. Perhaps the best example can be found in the period of the physiocrats, when it was believed that the productive capacity of the natural environment was the major source of welfare. However, other periods of history of economic thinking have paid less attention to nature as an important production factor. For instance, in classical economics capital and labour, in addition to land, were regarded as the main welfare generators. Furthermore, the classical economists assigned only a minor role to the government being an institution for establishing the framework within which market decisions have to be taken. However, it is interesting to note that also the classical economists were aware of the possibility of a stagnating economy caused by lack of natural resources.

As a consequence of neo-classical thinking, it was taken for granted in the post-war period that nature as such is not the source of welfare, but only the welfare constituents produced by labour, capital, technology and land. Clearly, land and nature have not become irrelevant, witness also the following quotation of Randall and Castle (1985, p. 573): "... there seemed no reason to accord land any special treatment that would suggest its role is quite distinct from that of the other factors. Land could safely be subsumed under the broader aggregate of capital,..."

After the neglect of environmental factors in Keynesian economics, we are the past decades facing a situation where the externalities and limits to growth (with regard to both renewable and non-renewable resources) have become a new focal point of economic research. The major question is, however, how to avoid a 'tragedy of the commons' (Hardin, 1968) in view of the long-term threats exerted by the (seemingly) inevitable and persistent changes in global environmental conditions.

In conclusion, despite a great deal of urgent environmental issues we still need a significant improvement of economic theorizing in this area. Admittedly, at a modest scale some progress has been made, but an operational methodology for long-term sustainability analysis is still missing. The contributions in the present volume have to be seen as attempts at coming to grips with the above mentioned issues. They reflect a wide spectrum of views and approaches and show that this field is still in motion. Various new conceptual frameworks are developed and attempts are made to devise an applicable methodology. However, a major problem in various contributions appears to be a clear view on the notion of sustainability (cf. Pezzey, 1989). This question will be further discussed in the next section.

Planning for Sustainability

Sustainability has become a key concept in current environmental debates. But the interpretation of this concept is less clear in many cases, as this may depend on underlying subjective or ideological views. In this context, Turner (1987) makes a distinction into four basic world views:

- extreme technocentrism: a resource exploitative, growth oriented position;
- accomodating technocentrism: a resource conservationist and managerial position;
- communalist ecocentrism: a resource preservationist position;
- extreme ecocentrism: an extreme preservationist position supported by an acceptance of bioethics.

Various reports presented in the past decade by different international institutions mirror - in one way or another - some of these (ideal-typical) views. The plurality in views and fragmentation in institutional responses however is an indication that a satisfacatory planning paradigm for sustainable development has not yet been found. Nevertheless, the world-wide support for the Brundtland Report indicates that a concensus on desirable future directions has in the mean time emerged on appropriate policy measures and controls (e.g., economic incentives, institutional mechanisms, etc.). In addition, there is a growing conviction that strategic policy thinking is a necessity, even though the precise meaning of sustainable development is not yet entirely clear. There is a clear need for including intertemporal effects (the multiple generations case) and trans-border effects (the echo effects case) in policy initiatives regarding natural resource degradation.

In general, there is a need for promoting policies that ensure a better resource management and conservation and, at the same time, reduce fiscal burdens on government and improve economic productivity. There is not by necessity a conflict between natural resource management and economic development policy, provided all market failures and distortions are adequately coped with. This holds also for countries heavily dependent on their natural resources for sustained economic growth (for instance, Myers, 1984; and Repetto and Gillis, 1988).

It seems to be a plausible assumption that sustainable development will not come into being unless all environmental sacrifices are fully reflected as costs to be charged to economic development. Planning for sustainable development means essentially a management of resources in which the direction and quality of environmental conditions are permanently monitored so as to have available full information for effective policy response. In this context a system of natural resource accounts might be devised in order to collect and store in a systematic way all relevant information on changes in the stock of (renewable and non-renewable) environmental assets. Such a system of natural resource accounts is essentially the dual side of a system of economic accounts monitoring economic developments. The way such environmental effects are measured (i.e., the economic valuation of environmental effects) is far from easy, but recent advances in the area of social cost-benefit analysis (notably marginal social opportunity cost analysis), multiple criteria evaluation and simulation modelling provide good examples of operational analytical tools. In this respect, also environmental impact assessment has to be mentioned, as this tool has proven its power in those countries where it has become an institutionalized part of environmental planning. A major challenge in the latter area will be to devise methodologies for resource accounting and environmental impact assessment that focus attention on the socio-economic dimensions of strategic sustainability policies.

Sustainability policies should be addressed to all actors involved: households, firms, national and international institutions, cities or regions, and government agencies. Information provision may be seen as a major task of sustainability policies at the micro level. On the other hand, there is also much scope for sustainability policies at the macro level, e.g., international agreements on a reduction of burning fossil fuel, cooperation through coherent multi-national environmental data banks.

In conclusion, planning for sustainability requires a shift in our thinking on the development of our economic system. There is an evident need for more strategic thinking, more cohesive thinking and more multidimensional thinking in order to ensure a compatibility of economic and environmental interests. This means that both the strategic significance of sustainable development and the implications for practical policy strategies have to be envisaged.

In the light of the above observations on sustainable development, the present volume is organized as follows. It starts - in part A - with a collection of refreshing and reflective contributions to the issue of a balanced economic and environmental development. The second part of the book, part B, is devoted to more practical and planning issues in this field, whilst here particular emphasis is placed on environmental assessment and policy evaluation.

References

Archibugi, F. (1974), A System of Models for the National Long-Term Planning Process, Report to the ECE(UN) Seminar on the Use of Systems of Models in Planning, Moscow, USSR, 2-11 December.

Barde, J.P., and Gerelli, E. (1977), Economies et Politiques de l'Environnement, Presses Universitaire de France, Paris.

Baumol, W.J. and Oates, W.E. (1975), The Theory of Environmental Policy, Prentice-Hall, Englewood Cliffs (New Jersey).

Boulding, K.E. (1966), The Economics of the Coming Spaceship Earth, in: Jarrett, H. (ed.), Quality in a Growing Economy, Johns Hopkins Press, Baltimore, pp. 3-14.

Bremer, S.A. et al. (1987), The Globus Model: Computer Simulation of Worldwide Political and Economic Developments, Campus Verlag, Frankfurt.

Ciriacy-Wantrup, S.V. (1952), Resource Conservation: Economies and Policies, University of California Press, Berkeley.

Clark, W.C. (1986), Sustainable Development of the Bioshpere: Themes for a Research Program, in: Clark, W.C. and Munn, R.E. (eds.), Sustainable Development of the Biosphere, Cambridge University Press, Cambridge, pp. 5-48.

Clark, W.C. and Munn, R.E. (eds.)(1986), Sustainable Development of the Biosphere, Cambridge University Press, Cambridge.

Dahlberg, K.A. (ed.) (1986), New Directions for Agricultural Research, Totowa, NY, Rowman and Allanheld.

Daly, H.E. (1973), Toward a Steady-State Economy, Freeman, San Franscisco.

Dasgupta, A.K. and Pearce, D.W. (1972), Cost-Benefit Analysis: Theory and Practice, MacMillan, London.

Doxiadis, C.A. (1968), Ekistics: An Introduction to the Science of Human Settlements, Oxford University Press, London.

Fox, K.A. (1971), Combining Economic and Non-Economic Objectives in Development Planning: Problems of Concept and Measurement, in: Sellekaerts, W. (ed.), Economic Development and Planning: Essays in Honour of Jan Tinbergen, MacMillan, London, pp. 141-158.

Fox, K.A. (1985), Social System Accounts: Linking Social and Economic Indicators through Tangible Behaviour Settings, Reidel, Dordrecht.

Georgescu-Roegen, N. (1971), The Entropy Law and the Economic Process, Harvard University Press, Cambridge (Mass.).

Georgescu-Roegen, N. (1973), Analisi Economico e Processo Economico, Sansoni, Firenze.

Hardin, J. (1968), The Tragedy of the Commons, Science, 13-12-1968, pp. 1243-1248.

Juster, F.T. et al. (1979), Social Accounting and Social Indicators: A Framework for the Analysis of Well-Being, Institute for Social Research, University of Michigan, Ann Arbor (Michigan).

Kneese, A.V. et al. (1970), Economics and the Environment: A Materials Balance Approach, Johns Hopkins Press, Baltimore.

Kapp, K.W. (1970), Environmental Disruption and Social Cost: A Challenge to Economics, Kyklos, vol. 23, pp. 41-68.

Leontief, W. (1970), Environmental Repercussions and the Economic Structure: An Input-Output Approach, in: The Review of Economics and Statistics, vol. 52, no. 3, August, pp. 262-271.

Leontief, W. (1973), National Income, Economic Structure, and Environmental Externalities, in: Moss, M. (ed.), The Measurement of Economic and Social Performance, National Bureau of Economic Research, New York, pp. 178-191.

Meadows, D.H. et al. (1972), The Limits to Growth, Potomac Associates/ Universe Books, New York.

Mesarovic, M.D. and Pestel, E. (1974), Mankind at the Turning Point, Dutton, New York.

Mesarovic, M.D. and Reisman, A. (eds.) (1972), Systems Approach and the City, North-Holland, Amsterdam.

Myers, N. (1984), The Primary Source, W.W. Norton, New York.

Net National Welfare Measurement Committee, Economic Council of Japan (1973), Measuring Net National Welfare of Japan, Tokyo.

Nijkamp, P. (1978), Theory and Application of Environmental Economics, North-Holland, Amsterdam.

Nijkamp, P., and Soeteman, F. (1988), Land Use, Economy and Ecology, Futures, vol. 20, no. 6, pp. 621-634.

Nordhaus, W.D. and Tobin, J. (1972), Is Growth Obsolete? National Bureau of Economic Research, Economic Growth, Columbia University Press, Columbia, pp. 3-14.

Norgaard, R.B. (1984), Co-evolutionary Development Potential, Land Economics, vol. 60, no. 2, pp. 160-173.

Norgaard, R.B. (1988), Sustainable Development: A Co-evolutionary View, Futures, vol. 20, no. 6, pp. 606-620.

Norton, B.G. (1986), On the Inherent Danger of Undervaluing Species, in: Norton, B.G. (ed.), The Preservation of Species, Princeton University Press, Princeton, NY.

OECD (1975), The Polluter Pays Principle: Definition, Analysis and Implementation, OECD, Paris.

Opschoor, J.B. (1987), Duurzaamheid en Verandering (Sustainability and Change), Inaugural Address, Dept. of Economics, Free University, Amsterdam.

Pearce, D.W. (ed.) (1978), The Valuation of Social Cost, Allen and Unwin, London.

Pearce, D. (1988), Economics, Equity and Sustainable Development, Futures, vol. 20, no. 6, pp. 598-605.

Pezzey, J. (1989), Economic Analysis of Sustainable Growth and Sustainable Development, The World Bank, Washington D.C.

Randall, A, and Castle, E.N. (1985), Land Resources and Land Markets, in: Kneese, A.V. and Sweeney, J.L. (eds.), Handbook of Natural Resource and Energy Economics, vol. II, North-Holland Publ. Co., Amsterdam, pp. 571-620.

Repetto, R., and Gillis, M. (eds.)(1988), Public Policies and the Misuse of the World's Forest Resources, Cambridge University Press, Cambridge.

Turner, R.K. (1987), Sustainable Global Futures, Futures, vol. 19, no. 5, pp. 574-582.

Wilkinson, R.G. (1973), Poverty and Progress: an Ecological Model of Economic Development, Methuen, London.

World Commission on Environment and Development (1987), Our Common Future, Oxford University Press, Oxford.

PART A

GROWTH, DEVELOPMENT AND SUSTAINABILITY

Chapter 2

'OUR COMMON FUTURE', SUSTAINING THE MOMENTUM

J. MacNeill

In February 1987 in Tokyo, the World Commission on Environment and Development (WCED) reached final agreement on the text of 'Our Common Future'. It had started 900 days earlier in Geneva. Its 23 members were then almost complete strangers - political and industrial leaders, scientists and diplomats - representing every shade of political ideology, from every region of the world, and with a strong majority from developing countries. It began in a state of almost total disagreement on many of the issues, but ended with a full concensus on a diagnosis of the issues, and on an agenda for change to manage the growing risks and the immense opportunities challenging the world community.

The response to 'Our Common Future' has been greater than anything one could have expected. It was on the agenda of the United Nations and its Specialized Agencies, of regional organization like OECD and the OAU Summit, and of many governments, national and local. Major corporations and industrial associations are reshaping their approaches to these critical issues in the light of it. So are a number of international and national policy institutes and non-governmental organizations. Universities are framing graduate courses around it. Schools are trying to find ways to build it into their curricula.

A Global Enquiry

I had the privilege of managing this global enquiry into the critical issues of environment and development. We did all of the normal things one would expect of a commission of enquiry. We commissioned expert papers, established high level panels, heard from distinguished world figures. But we also did something that was a world first: we organized open public hearings on every continent. As a result, we were able to meet and take evidence from thousands of people - from peasants to presidents, from indigenous peoples and slum dwellers to industrial leaders, and from local activists to the world's most prestigous learned societies.

'Our Common Future' was released in London in April 1987 and was presented to the United Nations on October 19, in a special plenary debate.

When the debate was scheduled, nobody knew that October 19th was to become better known as 'Black Monday', the day the New York Stock Exchange would drop by 502 points. That knocked us off the front pages, but is also underscored dramatically a recurring theme in the report - the need for some fundamental changes in the way we manage the planet.

Prime Minister Gro Harlem Brundtland, chairman of the commission, presented the recommendations. The Secretary General, Prime Ministers Ghandi, Mugabe and many other top political leaders spoke. Prime Minister Paul Schulter of Denmark spoke on behalf of European Communities. Several delegations included corporate leaders.

A dramatic highlight of the debate was provided by the President of the small island nation of the Maldives. He galvanized the Assembly with a plea for action to prevent the disappearance of his nation. Europe has seen many nations come and go as a result of man-made invasions of other men, but this is different. The President was talking about the physical disappearance of his nation resulting from a man-made invasion of nature.

The highest point in the Maldives is less than two meters above sea level. Most of the 1190 island nation is below one meter. A sea level rise of up to one and one-half meters within the next 30 to 40 years is highly probable, according to scientists closest to this issue.

To endangered species, we have now added a new category of global risks - endangered nations. The most densely populated and richest parts of many nations are on a fast countdown to extinction - unless preventive measures are taken. Given the inertia of the physical systems involved, and the time lags in the political systems involved, these measures - and measures to counter other global risks - must be initiated now.

A Global Emergency

There is no time to lose. We are living through an era of global emergency. We are winning the odd battle, it is true - the Montreal Protocol on Ozone is perhaps the latest example - but we are losing the wars.

On a global basis, the risks to the environment and for development have increased considerably during the past two decades. The battles won in the form of cleaner air, cleaner water, cleaner technologies, control of chemicals, more livable cities and international accords like the Law of the Sea have been real battles, but they have been limited in number and effect. They have also been limited to a few of the richer, industrialized nations.

During the past two decades of environmental action, the poorer countries of the Third World have witnessed a massive degradation of their environment and depletion of their resources - their bases for development. Forty years ago, Ehtiopia, for example, had a 30 percent forest cover; twelve years ago, it was down to 4 percent; today, it may be 1 percent. Seventy five years ago, India's forests covered over half the country. Today, they are down to 14 percent - and going fast. In the tropics today, ten trees being cut for every one planted -in Africa, the ratio is 29 to 1- forest cover is disappearing at a rate equal to the size of the United Kingdom every year. Water use is doubling in a decade. A mass extinction of species is underway. One square kilometer of one rain forest in one small country, Panama, contains more species than all of Britain.

An area larger than the African continent, and inhabited by more than 1 billion people, is now at risk from desertification and every year this area grows by 6 million hectares. The consequences include not only increased hunger and death but also social instability and conflict as dryland degradation drives environmental refugees in their millions across national borders.

Many countries in Africa, Asia and Latin America are today in deeper ecological debt than financial debt. But even the richer industrialized countries can no longer keep up with the pace of change. With a mounting tide of toxic substances entering the atmosphere, the waters, the soil and the food chain - between 1000 and 2000 new chemicals will enter the market place this year alone, most of them without adequate prior testing - with acidification accumulating in the environment, with the ozone layer threatened and with man-made climate change a looming prospect - with this and more, we are falling behind on just about every front.

All over the world, the Earth's basic capital of environment and renewable resrouces is being used faster than it can restore or replace itself. The world's potential for development is being reduced - in some areas to near zero - at the very moment in our history when, with population doubling, time collapsing, and human needs and aspirations exploding, that potential has to be increased. This cannot - will not - be sustained.

There is no doubt that our future survival is threatened, both as a world community of nations and as a species. The pressures which have generated these threats are bound to accelerate and, even if we start to act now, they are going to get worse before they begin to get better.

The Great Transition

We are living through a great transition in the relationship between the human world and the earth's biosphere, in the relationships between economy and ecology. The two have become interlocked irreversibly and, today, to ignore the one is to jeopardize the other.

One measure of this transition in **economic activity.** We are now in a $14 trillion dollar economy. Economic activity has increased more than 20-fold since 1900. The use of fossil fuels has grown by a factor of 30; industrial production by a factor of 50. And, four-fifths of all this growth has taken place in just 39 years, since 1950.

Another measure is **equity.** This economic activity has been concentrated in the industrial countries. With one quarter - soon 20 per cent - of the world's population, they consume about 80 per cent of the world's goods. That leaves three quarters of the world's population with less than one quarter of its wealth - an imbalance that has been getting worse, not better, but one that cannot be allowed to endure.

A third measure is the well-known **demographic momentum.** The world is now rushing towards a population of somewhere between 8 and 14 billion next century. Almost as many new people will be added to this Earth in the 4,500 days remaining in this century as existed in total at the beginning of it. Over 90 million will be added in this year 1988. Most of these children will be added to desperately poor families in the Third World - our neighbours in a global village.

In a world with the means and experience to eliminate it, mass poverty is not only a moral outrage, it is the greatest single failure of any society. As long as billions lead desperate and marginalized lives, their condition will be the greatest threat to the ecological basis of future survival and development.

A fourth measure is what we called the **growth imperative.** The world's economy must grow and grow fast if it is to meet the needs and aspirations of present and future generations. The Commission estimated that a 5 to 10-fold increase in economic activity would be needed over the next half century just to raise consumption in developing countries to more equitable levels. Energy use alone would have to increase by a factor of eight, just to bring developing countries, with their present populations, up to the level now prevailing in the industrial world. I could cite similar factors for food, water, shelter, and the other essentials of life.

Imagine what a five to ten fold increase in economic activity over the next half century means in terms of planetary investment in housing, transport, agriculture, industry, energy - you name it. Almost a new skin for planet Earth, certainly a new face. Talk about global change. It is unfolding before our eyes. We see more of it in a month than our forebears saw in a century.

Sustainable Development

The means and the experience exist to turn this apprehended crisis into a series of opportunities. We know, for example, how to gain 50 or 100 years on climate change, without sacrificing growth, deploying policies that have already been tested, while increasing corporate competitivity and macroeconomic efficiency. What more could we want in public policy? In fact, a number of trends and an enormous range of scientific and technological advances suggest that, with a little luck, we could turn the next century into a glorious renaissance - based on a cleaning and greening of the planet - and a greening of economics. That's what 'Our Common Future' is all about.

But it will require fundamental change. If we continue on our present course with 'business as usual', we will end up where we are headed.

'Our Common Future' is a political document. The main political concept of the report - that one that has caught the eye of governments, industrialists, scientists and others around the world - is sustainable development - that is, development which is both economically and ecologically sustainable. The report is full of suggestions about how to achieve sustainable development in the fields of international economic relations, energy, industry, agriculture, the management of the global commons, even new concepts of security.

A System Flawed

The changes occurring in our atmosphere, soils and water are spreading with an accelerating intensity that far outstrips the ability of our political and economic institutions to cope. These institutions are not working. Many believe they cannot work.

The essential problem is easily stated: the World's economy is totally interlocked with the Earth's ecology, but our institutions are not. Those institutions through which nations manage the environment are almost completely divorced from those through which they manage the economy. As a result, they are both increasingly ineffective.

Sustainable development requires strong measures to change both the content and the quality of growth and equally strong measures to ensure a more equitable distribution of the proceeds of growth, nationally and internationally. The decisions needed to achieve this cannot be taken by ministers of environment. Their decisions are usually focussed on the effects of development - health, property and ecosystem effects - and on reactive measures to deal with damage already done - measures to retrofit, re-build, replant, restore, reclaim, rehabilitate. These decisions are important and, given the trends, they will become even more important to our potential for development and survival. But they are not the decisions necessary for change.

No, the decisions needed now to turn the crisis around must be taken by ministers of finance, energy, agriculture, trade and so on - and by leaders of industry and science. In no country today have these ministers - whose policies are usually at the source of the accelerating drawdown of our environmental resource capital - been given any responsibility to take the sustainability of environmental resources into account. Nor are they ever held accountable politically for the impact of their policies and budgets on the sustainability of environmental resources. Both responsibility and accountability usually ends up on the shoulders of the poor and comparatively powerless minister of environment. Not an enviable position. Nor, given the imbalance of power and resources, a particularly hopeful one.

By and large, national environmental protection and resource management agencies, as well as international agencies, such as UNEP, have tried to do the best job possible, within the limits of their mandates and limited budgets. But nowhere have they been able to keep up with the growing pace and scale of environmental degradation.

And, looking back, it is unreasonable to have expected that they would. The central economic and key sectoral agencies of government, with their enormous policy and budgetary power, were and remain in a position to encourage forms of development that generate negative effects on resources and the environment at rates far in excess of the capacity of environmental and resource management agencies to deal with those effects.

Directions for Change

The Commission proposes several directions for change in these arrangements.

Clearly, we must strengthen our environmental protection and resource management agencies, both national and international, and multiply our spending on retrofitting, restoration and rehabilitation. The report contains a large number of specific recommendations in this regard.

The most urgent task, however, is to make sustainable development the overriding goal of all government agencies, national as well as international, including the entire UN system. Central economic, trade, and sectoral agencies must be made directly responsible for ensuring that their policies, and the budgets they command, encourage development that is sustainable, ecologically and economically.

Early Signs of Change

The task of securing an integration of environment and economics in our institutions will not be easy. But it can be done. And although our report appeared only a short time ago, much is happening.

The Commission recommended that the formal mandates of the UN Secretary General, the Specialized agencies, the World Bank, the regional banks and other regional organizations, should be changed to make sustainable development a primary goal of these bodies. The Commission also recommended a number of structural changes to better integrate environment and economics in decision making.

The resolution adopted following the General Assembly's debate, while necessarily general, endorses the general directions for institutional change recommended by the Commission. It calls for some action now and further consideration of our recommendations at the next session of the General Assembly.

Fortunately, a number of key agencies have decided not to wait upon the General Assembly. The World Bank, for example, is well into a re-organization and has greatly enhanced its environmental department so that it can better assess investment proposals against sustainability criteria. Some regional banks and several bilateral development assistance agencies, including CIDA in Canada, are taking similar measures.

The Commission proposed that parallel resolutions be introduced in the Assembly and later in the governing bodies of other agencies to make them directly responsible for ensuring that their programs and budgets support development that is sustainable. The governing bodies of two Specialized Agencies, FAO and UNIDO, have met. In both cases, member governments secured decisions requiring these agencies to review their programs in light of the Commission's recommendations, and to report back next year with proposals for change.

The Commission recommended a number of new international conventions, aimed at specific issues, including, for example, a Species Convention and a Convention on Climate Change. As a result, the Government of Canada decided to host a World Conference on 'The Changing Asmosphere' and its 'Implications for Global Security' in Toronto in June 1989. It is meant to achieve broad political support for policy action on atmospheric change, leading up to the 1990 World Climate Conference as well as the proposed 1992 World Conference on Sustainable Development endorsed by the General Assembly.

The Commission recommended that new and more dependable sources of financing be found to support programs to manage the commons, and to protect and restore basic earth resources and life support systems. As a first step in this direction, the Commission put forward the proposal for a World Conservation Bank, linked to the World Bank. It would provide loans and facilitate joint financing for the development and protection of critical habitats and ecosystems, including those of international significance. The recent agreement between Conservation International and the Bolivian Government under which the latter was enabled to write off $ 650,000 worth of debt to creditor banks for $100,000, or 15 cents on the dollar, in exchange for the establishment of a 3.5 million acre reserve of forests and grasslands in the headwaters of the Amazon, illustrates the potential for such a programme. The Commission urged that this proposal be subjected to an immediate feasibility study and this study is now underway.

A lot is happening at the national level, too, especially perhaps in the Nordic Countries, the Netherlands and Canada, but also in some other countries such as Indonesia. Let me mention Canada, where I come from.

When the Commission's report appeared in April 1987, a National Task Force was established to examine the relationships between Environment and the Economy. It took 'Our Common Future' and adapted it to the Canadian situation. The Task Force was made up of not only government ministers, but also the Chief Executive Officers of seven of our top corporations. It made a number of far-reaching recommendations for change in the way we do business in Canada and those recommendations were recently endorsed by the First Ministers of the country.

They will require major changes in the way governments approach the management of both the environment and the economy. They would require, for example, that the ministers responsible for the major productive sectors of the economy - finance, energy, industry, agriculture - be held accountable for the impact of their budgets on the environment. That is potentially very far-reaching, but even more far-reaching is their recommendation that all proposals for economic development policies and programmes that go to Canadian cabinets for decision should be required to demonstrate they are both economically and environmentally sound, and therefore sustainable.

The Task Force included corporate leaders. Among the many hopeful signs that some significant changes are already underway, none is more hopeful in my view than the changes which are evident in the attitude of many corporate leaders towards environmental protection and its relationship to economic productivity, competitiveness and growth. The confrontation built into older concepts of environment as a discretionary add-on to industrial investment is being replaced by an appreciation of the unbreakable links between a healthy industry and a healthy environment.

A key to real and effective strategies for sustainable development is increased co-operation between government and industry. The Commission recommended joint advisory councils in individual nations, and the National Task Force has proposed the establishment in Canada of what it calls 'Round Tables on Environment and the Economy'. These forums would consider the issues of sustainable development and policies to promote it on a continuing basis. Their recommendations would go directly to the Prime Minister.

The transition to sustainable development will involve policy and institutional choices that are politically difficult and the Commission felt that an immense effort will be required to inform the public and secure its support. Governments alone cannot provide the basis for making informed choices abount environment and development. The scientific community, industry, private and community groups and NGOs have a critical role to play in this.

The right to know and have access to information is fundamental to that role; so is the right to be consulted and participate in various ways. The right to have access to information is limited by the value of the information. We are currently flooded with data, but we have little in the way of information.

Information power is a reality. But no nation today has a strong capacity to provide timely and authoritative assessments and public reports on critical threats and risks to the community. Most have none at all. That must change and 'Our Common Future' contains a number of proposals for action.

The barriers to change, especially to fundamental change in the responsibilities and decision-processes of our major economic institutions, are enormous. The Commission and its Panels were made up largely of people with a strong political background. They were aware of these barriers and all the forms that they can take - ranging from absolute naivete to absolute corruption. They were also aware that, in the end, the bottom line was to find more effective ways to hold these institutions accountable.

There is, of course, an aresenal of techniques through which nations try to hold their institutions accountable: parliamentary committees, ngo's, independent policy institutes, auditors general, the media. There are not so many at the international level.

That is one reason why the Commission proposed that the General Assembly initiate negotiations on a new International Charter for Environment and Sustainable Development. Among other things, such a Charter would prescribe new norms to guide state and interstate behaviour in our global village. These norms would include, for example, prior notification to neighbouring states of activities likely to have an impact on them, as well as agreement to exchange information and to consult on request. Even after Chernobyl, the member states of IAEA were not prepared to go that far. They agreed that states had an obligation to alert and inform neighbouring states, but only after an accident has taken place that is likely to have a harmful impact on their environment. That's more 'react and cure' not 'anticipate and prevent'.

A great deal has happened in the recent past, but it will require an enormous and long-sustained effort to turn the Commission's recommendations into genuine action. With the presentation of the report to the General Assembly, the Commission completed its formal mandate and wound up its affairs as required by resolution 38/161. The responsibility for ensuring that its recommendations remain on the agenda and are addressed in an effective way now falls to the governments acting both within their own jurisdictions and internationally through the UN General Assembly, regional organizations like ASEAN, and OAU, COMICON and the OECD, the banks and assistance agencies. Industry, the scientific community and non-governmental organizations have critical roles to play to maintain the momentum for change.

The Commission proposed and the General Assembly has adopted a series of benchmarks for progress in the transition to sustainable development. A number of meetings and conferences are planned including, we hope, a series of regional conferences on sustainable development in 1990. Norway and Canada have already announced such conferences. Others are being considered. The General Assembly gave the green light to a World Conference on Sustainable Development in 1992 to review progress made and promote further follow-up arrangements to set new benchmarks that will be needed to maintain human progress within a framework that is globally sustainable. Canada and Sweden have both offered to host it.

Beyond that, there are a series of meetings planned in each of the sectors where decisions critical to sustainable development need to be fostered - or taken. I am referring to the annual meetings of the Economic

Summit, the Region Summits and the Banks, the ministerials of OECD, ASEAN and other regional associations; the Inter-Parliamentary Union. I am also referring to specific events such as the forthcoming World Climate Conference. Several major non-governmental organizations, scientific and industrial bodies, and policy institutes are developing - and some have launched - programmes and events, to advance consideration of the report and support for its recommendations.

The General Assembly, we hope, will decide to initiate a process to transform the report into a UN Programme for Sustainable Development. That would stimulate a wide range of activities for change leading up to the regional conferences in 1990, the World Conference on Sustainable Development in 1992 and other meetings.

Commissioners, of course, will continue to play a leadership role in these activities to the extent that their official and private duties permit. Some of us have felt for some time that there is a need for a small center to act as a focal point for these efforts over the coming years. The Secretariat, along with some others, has been working to establish such a focal point and as a result a 'Center for Our Common Future' will be established on April 1, 1988. It will be centered in Geneva.

As difficult as it may be, our political and policy systems must provide the leadership needed to secure a transition to sustainable development. If such leadership is provided, people will respond. 'Our Common Future' has uncovered a deep mood for change, and acceptance of the fact that change must come, and that now is the time. We see it in the polls - since 1978 a growing majority in all OECD countries have supported stronger measures to manage the environment.

The forces which are shaping the future of our planet today are more than ever the product of human activity and subject to human control. We are literally in command of our own future on this planet and whether we succeed or fail, the decisions and the responsibility will have been ours. Europe, following its Year of Environment, is better placed than most to provide leadership and influence the decisions that are needed.

The challenge is immense. To succeed we will need to apply a new way of thinking, a new mind-set that gives priority to the integrity, stability and beauty of the Earth we have inherited, and that reflects a commitment to uphold the right of all peoples and nations to participate equitably in the resources of this planet and the benefits of the development processes that they support.

Chapter 3

ECOLOGICAL SUSTAINABILITY AND ECONOMIC DEVELOPMENT

D.E. James

P. Nijkamp

J.B. Opschoor

Introduction

Sustainable development is a concept recently proposed as a guiding principle for economic development planning (IUCN 1980, WCED 1987). It is applicable in both developed and developing countries. The most widely promulgated definition of sustainable development is that given by the World Commission on Environment and Development: development is sustainable if it satisfies present needs without compromising the ability of future generations to meet their own needs (WCED 1987). Goodland and Ledec (1986) define sustainable development as "a pattern of social and structural economic transformations which optimises the benefits available in the present without jeopardising the likely potential for similar benefits in the future".

In the above definitions, there is no direct or exclusive reference to the ecological viability of particular patterns of development. Implicitly, however, a major ingredient in the concept of sustainable development is the need to maintain yields from renewable natural resources over long periods of time. Ecosystems can thus be expected to play an important role in policies and planning aimed at achievement of the goal of sustainable development.

This chapter gives a fuller interpretation of the concept of sustainable development, noting the economic significance of the biosphere in providing part of the economy's capital stock. From an economic viewpoint, key questions that must be answered include:

- what is the optimal size and composition of the economy's environment capital stock?
- how successfully can the tools of economic analysis be applied to environmental management?
- how should the problems of risk and uncertainty in ecological management be handled?
- to what extent should goals other than economic efficiency be taken

into account when making decisions on exploitation rates and preservation constraints in natural resource decisions?

Two Perspectives on Sustainability

The emphasis in the two definitions of sustainability given above is on the phenomenon of development or growth, which is to be continued. The WCED does refer explicitly to the limitations of the environment to meet present and future needs, but these depend partly on the state of technology and social organisation. The Report states that "in essence, sustainable development is a process of change in which the exploitation of resources, the direction of investments, the orientation of technological development, and institutional change are all in harmony and enhance both current and future potential to meet human needs and aspirations" (WCED p. 46). This is essentially an economist's approach to the concept of sustainable development. It is the level of welfare that is to be sustained or perpetuated through economic, institutional and technical change.

Other approaches to the concept of sustainable development focus on the physical or natural resource base of an economy. According to some economists, sustainable development implies maintenance over time of aggregate resource stocks, such that the potential to generate welfare is not permitted to fall below the current level (Pearce 1986, Jacobs 1985, Repetto 1986). The World Conservation Strategy (IUCN 1980) pleaded for utilization of the biosphere in such a way that life support systems, essential ecological processes and species diversity be maintained and improved. Such an approach might be labelled the environmental approach, in contrast to that of the economist.

Environment as Infrastructure

Underlying the concern over sustainability is the notion that the environment provides an infrastructure without which the economy could not survive. All materials and energy transformed by production activities into economic goods and services originate from nature. For flows of materials and energy to continue, ecological systems and processes must remain in operation and may also need to be reinforced. Biogeochemical cycles should continue to circulate materials in the biosphere, ecosystems should retain their capacities for the assimilation and degradation of wastes, and renwable resources such as fish populations, forests, and soils should maintain their regenerative potential. These may be referred to as the "productive infrastructrual functions" of the biosphere. In simple terms, they represent the economy's "storeroom".

Apart from these productive functions, the biosphere serves as a habitat for Homo Sapiens. It provides shelter, including protection from radiation, and amenities such as landscapes, scenery and biotic diversity, that add to human welfare. These can be called the "consumptive infrastructural functions" of the environment. In this capacity, the biosphere can be said to represent the "living room".

The biosphere is also host to a large number of species apart from Homo Sapiens. There may be a connection with human welfare in that satisfaction is derived from the existence of larger species diversity, but here is meant something much wider, namely the provision of a life support system for other species. The greater the intrusion of economic activities into the biosphere, the less hospitable becomes the environment for these species. It is estimated that, at present, human activities are responsible for species extinction at a rate 10 to 100 times the natural rate.

Below, we focus on the infrastructural functions of the biosphere and assume that its "integrity", as indicated by species diversity and the size of biotypes, is maintained at acceptable levels. Given those levels, the patterns and levels of economic activities and their environmental impacts, and also given the prevailing state of technology, the environment provides a range of resources, including capacities and services, that form a multidimensional "resource utilisation space" (Siebert 1982, Opschoor 1987). Some of these resources do not diminish as they are consumed (and are thus described as public goods), but most do. Other resources serve demands that are non-essential (thus constituting luxury goods). Many, however, provide the infrastructure for the production of goods and services to satisfy basic needs. Welfare and development depend strongly on this last category of natural resources. The continued functioning of natural processes to support such requirements takes place within structures described here as ecosystems. In the discussion that follows, we concentrate on those parts of the productive and consumptive environmental infrastructure that relate to essential resources, materials and services.

Environmental Capital and Welfare

Welfare depends on the availability over time of an adequate environmental utilisation space. What economist used to call "land" consists of a black box that is better referred to as "environmental capital". Looking only at the outputs from the box as resource flows or yields, the box contains a set of "resource regeneration systems". Environmental capital would then be conceived as an aggregate of all individual resource stocks and their regeneration systems.

Welfare depends on more than environmental capital. Other preconditions include a man-made physical infrastructure, a productivity-related stock of knowledge and skills and an institutional infrastructure governing decisions affecting welfare. Classical economics concentrated on the welfare implications of land, labour and man-made capital, and neoclassical economics on man-made capital and human capital. Neoclassical environmental economics has reintroduced land as environmental capital, although essentially as a single-output, single process system. A growing number of economists (Kapp 1974, Norgaard 1984, Daly 1973) approach society-environment interactions from an integrated systems viewpoint, attempting to provide more detailed, realistic analyses of the environmental infrastructure, assess the significance of the institutional structure and make suggestions for improving it.

Welfare depends on all types of capital. Physical, environmental and human captial are all interdependent and can, to a certain degree, serve as substitutes for each other in terms of providing flows of production factors into a welfare generating process.

Neoclassical economists in particular rely on these substitution possibilities and on the role of technology to maintain welfare levels. Solow (1986) therefore holds that the present generation does not so much owe future generations a certain share in the stock of environmental capital, but rather: "access to a certain standard of living or level of consumption" regardless of the form in which this is bestowed. Such a view, however, could comprise a blank cheque on future technological possibilities and hence a laissez-passer for the overexploitation of natural resources. It also disregards the hierarchical dependence of one form of capital on another. Man-made capital, for example, is bounded by the availability of environmental capital, and human capital may be bounded by both forms of capital. In fact, only environmental capital has, in its (semi) renewable resources and regeneration systems, an autonomous base for extended existence.

It is fallacious to suggest that human welfare has become less dependent on environmental capital. Mankind is drawing more heavily than ever on environmental resources and is already responsible for large overdrafts on the environmental capital account. Anticipated population growth and rising per capita real income over the next few decades imply even heavier demands on the environment. It can be shown that even if the material content of real income is reduced by 80 per cent over the next six decades, the global impact of human activity on the environment will more than double. Worse still, such development may trigger irreversible trends towards species extinction and ecosystem destruction on a scale never previously seen.

Two important counteracting dynamic factors influencing the size of the environmental capital stock should be noted. The first is possible enhancement of the productivity or welfare generating capacity of environmental capital, especially the further evolution of human knowledge and skills, the discovery of new resources and the development of new ways of using existing resources. Each "unit" of environmental capital may thus yield more welfare, or conversely, the same level of welfare may be reached with a smaller input of environmental capital. Good examples are energy conservation, recycling, and materials substitution. The second factor is the indirect and long-term effect of degradation of environmental capital on regenerative systems and buffering processes comprising the environmental infrastructure. Such mechanisms often involve long time lags between initiating actions and subsequent responses. Actions or decisions taken now may significantly reduce the environmental utilisation space in the future.

Finally, the influence of institutional structures should be considered. The use of environmental resources depends on the formal and informal organisation of society and on prevailing social values. Any alteration of institutional patterns may lead to an alteration of activity patterns with respect to the use of environmental capital. Reductions in environmental impacts can be achieved through improved environmental awareness, new instruments of environmental policy, and more powerful national and international institutions for curbing environmental degradation.

Unless technological innovation and improvements in national and international resource management are greatly accelerated, the environmental utilisation space is in serious danger of shrinking rapidly.

Valuation of Environmental Capital

The total value of the environmental capital stock in principle can be measured as the present value of future net benefits derived from use of the stock in future years. But whose values should be used to achieve such a measure? Should individual values be weighted according to income distribution (the willingness and ability to pay) or should some more egalitarian set of weights be adopted? Should the measure depend only on the values of present generations or should those of future generations somehow be taken into account? If the latter are important, what time horizon should be chosen, and how much weight should be given to each future generation? Answers to these questions by definition imply a social value judgement regarding the distribution of resources over time.

A second problem to be addressed is determining which functions of the environment have value for individuals. Typically, these consist of user benefits and non-user benefits. The user benefits are composed of the net market value of environmental attributes, consumers surplus, and the value that the individual places on retaining the option for use of the environment in the future. The non-user benefits reflect the individual's preference for leaving certain environmental attributes for future generations (bequest value) and an estimate of the "existence value" or intrinsic value of preserving such attributes, even if there is no associated use value. Purely functionalist and egoistic individuals would count only user benefits. Bequest values reflect individuals positions in relation to the preferred level of intergenerational solidarity. Existence values reflect individuals' preferences towards nature conservation. It is apparent that non-user values are difficult to determine empirically. However, they also involve more fundamental problems. For example, society as a whole may take a different view on bequest values, with a different solidarity weighting; and societal values may also differ for existence values or even for user values on "merit" grounds (Opschoor 1974, James et al, 1978, Siebert 1987). This simply means that individuals may not be accepted as the sole and best judges of societal values of the environmental infrastructure. Societies taking this view must derive their values on the basis of merit considerations.

Finally, if societal values are based on a simple aggregation of individual monetary values, a rather strong value judgement is made. One of the critiques of the Neoclassical economic approach, for example by the institutionalist or evolutionary economists, is that such an assumption is unwarranted. People and societies typically have value structures and value hierarchies, which means that ethical positions taken at one level of values may pose restrictions on the range allowed for other values. In terms of environmental management, this may imply that rather than basing decisions on calculated functions of solidarity or merit values, use of the environmental capital stock is physically restricted. If this approach is preferred by society, then the rate of use or depletion of a single resource or of the entire environmental capital stock, cannot be determined on economic efficiency grounds alone. Simple comparisons of benefits and costs are inappropriate under such circumstances.

The Optimal Level of Environmental Infrastructure

To ensure that future generations' needs can at least be met at today's levels, it is necessary to decide how much environmental capital should be preserved and what its composition should be. The following factors should be considered:

- ethical notions such as the level of solidarity owed to future generations;
- the present and future degree of (riskless) substitution between the various elements of the environmental capital stock;
- the present and future degree of (riskless) substitution between elements of environmental capital, man-made capital and human capital;
- the degree of robustness of the biosphere in accommodating different spatial distributions of economic activities and the associated environmental impacts;
- the future institutional capacities for managing or adapting to environmental changes.

It is assumed here that rising material welfare claims can be expected from a rapidly rising world population and that this should be of moral concern of societies today. Assuming this, and also taking a risk-averse approach to the future role of science and technology, a policy of conserving the present environmental infrastructure can be advocated. Such a policy might, for practical reasons, be initially based on arbitrary judgements about the requirements for sustainable development. Ideally, judgements would be supplanted by more sophisticated analyses addressing interdependencies within the environmental infrastructure, projections of technological and institutional change, population growth, and likely patterns of economic inequality on a global scale.

Risk, Uncertainty and Option Theory

Care of the biosphere cannot be entrusted to a hidden hand, governing market operations. Human interventions in the area of land use, technology and resource extraction are often a threat to ecologically sustainable economic development.

A socially acceptable balance between long-term sustainability and development is difficult to identify in the absence of an agency which sees its explicit task as preservation of the environment for future generations (cf. Davis, 1987). Such a task is made difficult by restricted insight into the effects of current economic, technological and physical planning on future environmental quality. Thus, environmental and resource management is a typical example of risk behaviour.

This situation is comparable to portfolio problems, where the goal is also fairly well defined but where the selection strategies are hampered by lack of information on future events. In the recent literature on this issue, reference is often made to the relevance of option theory. Option theory was originally developed for evaluating the effects of uncertain future options on corporate stocks, but it can also be applied to other choice problems under uncertainties, e.g., in investment in R & D or in natural resource investment planning (Mason and Merton, 1984). In such cases the option is to enlarge or reduce the scale of a project, to abandon the project or to look for a sequential approach.

Risk and Uncertainty in Environmental Management

Environmental policy is essentially also a risk strategy which serves to minimize the mismatch between economic development and ecological sustainability under uncertain future conditions. In a way analogous to business behaviour one could assume that an environmental management agency is a specific type of entrepreneur whose final goal is to maximize ecologically sustainable economic development. In more concrete terms, one might for instance assume that this entrepreneur attempts to maintain the long-run value of the stock of "environmental capital", whilst also ensuring sufficient ecological diversity if only to minimize the risk of instability in ecosystems. The latter strategy is also regarded as meaningful in portfolio selection problems. Such environmental proservation strategies have been advocated several times in the economics literature (Hyman and Schmidt, 1983; Weisbrod, 1964).

Clearly, an "environmental entrepreneur" has to operate in an institutional setting with many conflicting objectives (from both the private and the public sector), but this situation is no different from normal business practice, which is also characterized by highly competitive behaviour under risk conditions.

An environmental entrepreneur must be aware of economic conditions that could impact on his capital stock. In a period of economic decline, environmental interests tend to receive a lower rank order on a societal preference scale. In a period of economic prosperity, increases in production and supplies of commodities will adversely affect the natural resource base. The environmental entrepreneur may thus be engaged in a permanent fight with economic entrepreneurs to guarantee ecological sustainability. Fortunately, however, economic well-being in the long run is dependent on the quality of the biosphere. Thus in the long run, the environmental entrepreneur should be an ally of economic entrepreneurs. This convergence of interests from a long-term perspective is likely to exist in various sectors of environmental management and land use. For instance, Nijkamp and Soeteman (1988) state that agricultural land use can in the long run be sustained only when farmers do not destroy the environmental conditions that provide the resource base for agriculture.

Coping with Risk and Uncertainty

The fact that conventional economic inputs and environmental capital both may contribute to welfare and may involve a trade-off could imply the need to impose minimum achievement levels (or threshold levels) for both economic and environmental conditions, in order to survive. For instance, Ciriacy-Wantrup (1952) several decades ago advocated the use of a minimum bequest value in strategic environmental policies, in particular the establishment of safe minimum standards of conservation to avoid critical zones beyond which it becomes uneconomical to halt and to reverse depletion. Thus planning for co-evolutionary development (Norgaard, 1984) has to be pursued with careful consideration of threshold levels for both the economic and environmental systems.

In this context, the long reaction times of economic and ecological processes often conceal the inexorable links between past and future events as evident in such phenomena as deforestation, ozonization and desertification (Crutzen and Graedel, 1986; Dahlberg, 1986). Thus the environmental entrepreneur may face sudden surprises in ecological transitions, including ecological "time bombs". The sudden death of thousands of monk seals in the North Sea in 1988 is a clear illustration of this phenomenon.

A revealing example of an ecological surprise is Big Moose Lake in New York (NAS, 1986). As shown in Figure 1, the lake had a fairly stable pH level until the 1950s. From the 1920s on, however, the lake was subjected to extremely high levels of acid deposition. Figure 1 shows the abrupt increase in the pH level almost 30 years after the acid deposition started to become very severe.

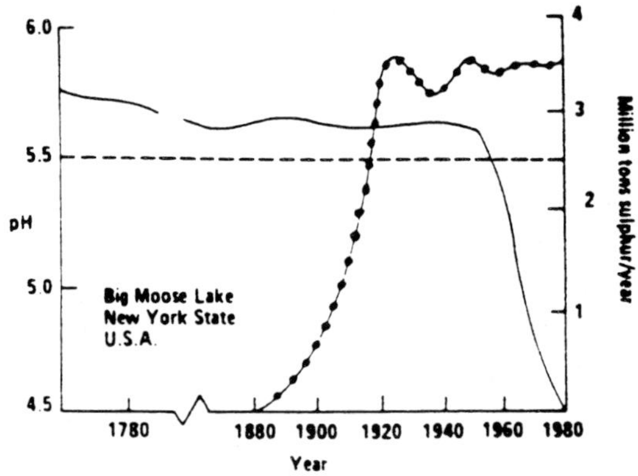

Figure 1. Long-run evolution of the pH and of sulphur in Big Moose Lake

Strategic Environmental Action Plans

Manning (1988), has developed a pyramidal framework which may be regarded as a strategic action plan for an environmental entrepreneur. He distinguished the following levels:

- goal of sustainable development: maintaining the resource base (land, water, air) to sustain land and economic activity;
- implementation: activities related to environmental resource base, which are needed in order to achieve the goal (e.g., planning and management of resources);
- development of solutions: establishing policy objectives, development of mitigation methods and of protecting techniques;
- evaluation of problems: impact analysis, evaluation of resource allocation problems and of trends, etc.;
- analysis of facts/trends: design of forecasting methods, development of scenarios, analysis of constraints and carrying capacity, development of assessment techniques, etc.;
- data collection: socio-economic, demographic, resource use, environmental and biophysical data, etc.

The Role of Information

In the light of uncertainties surrounding sustainability strategies, a prudent environmental entrepreneur would need more specific information to make balanced economic-environmental decisions. Such information might not take the form of quantitative data, but rather a set of guiding principles.

An environmental entrepreneur who is facing the above mentioned uncertainties in a long-term perspective would base his strategies on the following:

- identification of the **crucial ecological processes, ecosystems and resource regenerative systems** which form the infrastructure of the "natural capital" from which welfare is to be derived in future (WCS, 1982; Opschoor, 1987), and of an appropriate set of indicators to monitor quantitative and qualitative changes in these processes and systems (environmental indicators on regulatory, carrier, production and information functions of the environmental capital; resource accounting; see De Groot (1987);
- identification of **long-run driving forces** which impact on both the economy and the biosphere (e.g., demographic transitions, technological developments, socio-cultural changes) and which will have a supra-marginal influence on the future welfare of society through sometimes complex economic-environmental interactions. In practical terms this means identification of variables that are most critical for long-run sustainability (see for an illustration Galeski and Wilkening, 1987);
- identification of conditions under which **perturbations** (shocks, discontinuities, surprises, etc.) may come about; imaginative thinking, creative environmental entrepreneurship and expert knowledge are necessary conditions for becoming aware - in time - of hardly foreseeable events (see for an example, NAS, 1986; Manning, 1988; Holling, 1986). Hazard models, event history analysis and chaos theory, may be meaningful tools here (Mandelbrot, 1977; Nijkamp and Reggiani, 1988). Surprise management is perhaps the key issue in environmental sustainability policies, as this is based on anticipatory attitudes that induce preventive actions or rapid adjustments;
- identification of **irreversible evolutionary processes** (e.g., the transformation of forestry into urban land). The aggregate result of numerous small-scale land use transformations may be devastating to environmental quality conditions in the long run, as there is no scope for reversibility in the foreseeable future (Randall and Castle, 1985). Irreversible processes may represent a serious threat to an ecologically sustainable economic development, as they reduce the degrees of freedom (future options) available to an environmental entrepreneur;
- identification of long-term **feasible** (technical, economic, demographic, social and ecological) **boundaries** within which economic and

environmental developments can take place. Clearly, the demarcation of feasible decision spaces for environmental entrepreneurs is not an easy matter (Opschoor, 1987), although recently, in a comparative static context, some interesting attempts have been made (WRR, 1988).

Some Broader Considerations

The concept of an environmental entrepreneur is clearly not a realistic one. It is comparable to the notion of a policy-maker in macro-economics, which is also a fictitious concept. Nevertheless the personification does help to explain the way in which decisions are prepared, considered and made. In a specific context, institutional arrangements would have to be spelled out in greater detail. However, some rational guidelines have been derived.

From the strategic viewpoint of an environmental entrepreneur-megatrend analysis at a meso level focussing on qualitative changes and major directions of influence and change - may be more important than seemingly precise model predictions. The environmental entrepreneur could use the results from plausible strategic long-term scenarios, expert views and megatrend analysis as a frame of reference. Special attention should be given to development paths that diverge from conventional wisdom. The entrepreneur would also benefit from continuous monitoring of environmental transformation supported, for example, by strategic natural resource accounting and early warning systems based on geographical information systems.

Application of Economic Techniques to Ecosystem Management

As shown in the earlier part of this chapter, achieving sustainable development is a multiple-objective exercise, involving economic efficiency analysis, risk assessment and ethical judgements on the distribution of benefits and costs, both currently and with respect to future generations. To achieve long-term sustainable development, economic activities must be compatible with the functions of the biosphere. Economic techniques must be capable of addressing this policy objective.

The new emphasis on renewable resource productivity implicit in the concept of sustainable development gives a standpoint from which to address resource management problems, assessing options and consequences, and deriving workable recommendations and guidelines for policymakers and resource managers.

At present there are significant gaps between the standard theory of natural resource management and the economic assessment techniques that are applied in practice. The management of natural resources can be approached at different levels, ranging from broad macro and sectoral development policies to management at the level of the resource user. Most practical applications of economic analysis are made at the resource user level. In such a setting, the role of economics in implementing the concept of sustainable development is to seek a rationale for resource user behaviour, analyse the associated benefits and costs, assess the welfare effects with respect to resource users themselves and to society as a whole, and recommend appropriate rectification incentives, rules or procedures that will enhance community living standards. Where there is a divergence of interest between resource users and society as a whole, governmental intervention or public assistance of some kind is required.

New tools of economic analysis are being developed for application to natural resource management, and existing tools are being refined and used more widely. Benefit-cost analysis is perhaps the most commonly used tool for economic efficiency analysis, especially when used in combination with ecological simulation modelling. Better physical data systems and environmental impact assessments will facilitate the correct measurement of environmental benefits and costs. Much of the uncertainty surrounding estimates of the costs and benefits of development is the result of inadequate scientific analysis rather than economic valuation procedures. A multidisciplinary approach is essential.

There is a clear need to develop better accounting methods to record more accurately social benefits and costs of impacts on the natural environment and their links with economic activities, assess the true social rate of return on man-made capital at the national level and the social rate of productivity for natural resources. These are major aims (although not the only aims) of natural resource accounting.

Economic assessment methods, despite their recent improvements, are far from perfect, nor are they always sufficient from a decision-making point of view; the "merit"-argument developed above, underlines this. In such cases, hybrid approaches to resource management are necessary. The kinds of assessments and recommendations required to achieve sustainable development may not satisfy economic purists, but practical measures are urgently required to correct some of the more glaringly obvious and often life-threatening breakdowns in renewable resource management that are currently occurring throughout the world.

Applications of Economic Efficiency Analysis

Although other social goals may be important in the management of natural resources, the potential contribution of economic efficiency analysis to better use of the environmental infrastructure should not be overlooked. More effective application of benefit-cost analysis should result in improved resource management. The theory is well developed. Social economic benefits of resource use should be assessed in terms of multiple use, not necessarily revealed in market prices and behaviour. In principle, existence and bequest values can be incorporated in the analysis. Social economic costs of natural resource use should be measured in terms of marginal opportunity cost, which contains the following elements:

1. direct costs of extraction or harvesting;
2. external costs; and
3. user costs, i.e., the future benefit forgone by harvesting now rather than later.

In a world of perfect information, economically efficient use of resources could be achieved by the routine application of benefit-cost analysis. The usual economic aim is to identify the management option that maiximises NPV (Howe, 1979). Simulation modelling accompanied by benefit-cost analysis will automatically incorporate the marginal opportunity cost approach. The imaginative exploration of resource management alternatives in economic-ecologic simulation models is a relatively straightforward exercise. Through the generation of net benefit response surfaces, economically efficient resource management strategies can be identified (Holling, 1979).

Natural resource productivity, especially on a regional scale, can be greatly enhanced by better assessment and internalisation of spatial external environmental effects. Considerable progress has been made in this area. Case studies in the Asia-Pacific region are reported by Dixon and Hufschmidt (1986), among others. Improved methods of synthesising applied science and economics are strengthening the empirical base to facilitate resource management decisions.

Much is made of the failure of economics, through the use of discount rates and limited planning horizons, to provide for future generations. In practice, resources are often abused because resource users themselves adopt only a short planning horizon or high discount rate. The welfare at stake is often that of present generations, let alone that of future generations. Under extreme circumstances (e.g., during famines or where resource use rights are poorly defined) the private rate of time preference may exceed the social rate, warranting corrective public action. There is fairly widespread acceptance by economists that the social discount rate should be lower than the private rate. Resource degradation can be prevented by inducing resource managers to extend their planning horizons to those conventionally used in economic analysis or to use a lower discount rate (Dixon et al., forthcoming). Alternatively, more directly, various types of ecological stock constraints might be imposed.

Limitations of Benefit-Cost Analysis

Remedial policies clearly involve significant shifts in scarcity prices (Pearce, 1988). Is it possible to predict the "optimal" set of prices? And can the social "optimum" be approached on a piecemeal basis? The theory of second best has major implications here. Perhaps the only hope is to persist in the application of conventional tools of analysis, incorporating the broadest possible horizons, both spatially and temporally.

Conflict between economic objectives and ecological preservation reaches a maximum when economic efficiency analysis recommends extinction of natural capital stocks. This will be the case where the rate of growth of net benefits from renewable resources is less than the discount rate. It will then be economically efficient to reduce the stock to zero and invest the proceeds in man-made capital from which, as indicated by the discount rate, higher sustainable increases in national income can be expected. Sustainable levels of real income can be promoted, that is, by converting ecological capital to man-made capital which has greater productivity. This argument requires that net benefits have been correctly estimated and that running down a particular resource is compatible with sustainable use of other resources.

Guidelines for Ecological Preservation

As a result of obstacles in applying the techniques of economic efficiency analysis, and because of the need to accommodate objectives other than economic efficiency in achieving sustainable development, hybrid approaches to decision in the management of natural resources may be the only practical solution. The principal guidelines suggested by different economists include:

1. lowering the discount rate;
2. imposing renewable resource stock constraints;
3. using environmental indicators and standards;

4. enforcing resource protection, restoration and enhancement.

Such guidelines should be implemented with great care. It is important not to overemphasise environmental conservation, even though there may be deficiencies in economic assessment techniques and information. The possible influence of politicians, lobby groups and extremists must be resisted. The best available scientific advice is needed. Economic efficiency analysis can assist the critical appraisal of resource management options. There is a long history of such experience in the field of pollution control, where an inability to clearly specify net benefits of control programs has led to the imposition of standards, based largely on scientific opinion.

Lowering the Discount Rate

The approach to discounting conventionally recommended in natural resource management is to apply the same discount rate as in all other investment opportunities in the economy. This practice has been recently questioned (Gijsbers and Nijkamp, 1988).

Stocks of renewable resources bequeathed to future consumers will tend to vary inversely with the discount rate, although steadystate stocks (and yields) can be maintained even for a wide range of discount rates (Clark, 1976). If a high discount rate is applied in natural resource management, renewable resources will tend to be exploited at higher rates, leading to a lower stock size or to extinction of the stock at the end of the planning period, depending on the alternative uses of the resource with their respective net benefits, and the discount rate that is applied. As a consequence, the economic welfare of future consumers may be adversely affected.

Use of the standard discount rate in resource management effectively compares the rate of return on man-made capital with the rate of return on natural capital. This rate depends heavily on the internal rate of return on investments in man-made capital.

The accepted rate of discount could, however, be overstated for the following reason. Even though the reliability of economic assessment techniques has improved markedly in recent years, it is virtually impossible to capture all the environmental degradation costs of investments in man-made capital. The true costs are hidden by the tyranny of (relatively) small decisions. In general, because of the failure to accurately measure the true social costs of man-made investments, the estimated social rate of return will be overstated.

By contrast, the benefits of natural resource stock preservation tend to be underestimated. Natural resource use may in some instances lead to external costs that are difficult to measure, but the weight of ecological evidence suggests that, in general, significant external benefits are associated with conservation of natural capital stocks.

A case can thus be made for lowering the discount rate on man-made capital when transferring it to the management of natural resources, and increasing the implied relative rate of return on natural resources. This will tend to result in the maintenance of larger stocks of natural resources and create more options for present and future consumers.

The notion of lowering the discount rate in natural resource management to achieve larger stocks of renewable resources is, nevertheless, only a crude substitute for more precise measurement of external benefits and costs. The more accurate such measurements become, the weaker will be the case for lowering the discount rate, at least on economic efficiency grounds. Stock preservation may, of course, be advocated on grounds other than economic efficiency, as previously indicated.

Imposition of Ecological Stock Constraints

Preservation of ecological stocks at the end of the planning period has been advocated as a means of achieving the multiple objectives associated with sustainable development (Pearce and Markandya, 1987; Goodland and Ledec, 1987). This can be achieved by controlling exploitation rates to preserve the original stock, incorporating stock protection and rehabilitation measures, and introducing special programs and projects to restore the resource stock at the end of the planning period.

A problem in reality is deciding exactly what should be preserved. Is it a natural capital stock identical in size and condition to the stock at the beginning of the planning period, a stock that is capable of delivering the same sustainable yields beyond the planning horizon, or a stock that may be depleted but which maintains the potential for self-regeneration or rehabilitation by future generations?

Sustainable yields from natural capital can in many cases be increased by reducing the size of the stock, thus the original stock size may not be the optimal stock to bequeath future users. In other instances, sustainable yields may be increased by increasing the size of the stock. In either event, the benefits and costs of different levels of yield are what matter rather than the size of the stock in itself.

Environmental Indicators, Critical Zones and Safe Minimum Standards

The use of ecological indicator variables to prevent ecosystem deterioration is well established in the resource economics literature. Ciriacy-Wantrup (1952) introduced the concept of critical zones long ago, and advocated the use of "safe minimum standards" to preserve stocks of renewable resources. The concept has been further explored by Bishop (1978).

There are various difficulties in applying the concept of SMS. The first is establishing the basis for selecting appropriate indicators. The links between indicators and ecosystem performance must be well documented and understood by resource managers. Information is also required on tradeoffs between economic gains of ecosystem exploitation and the risks of ecosystem damage, especially in critical zones. Walters (1986) ilntroduces the notion of "surfing" in discussing this issue.

More important is the way in which the SMS is implemented, especially in a dynamic management context. Should a time-invariant management rule be established, or can the rule be made sufficiently flexible to adapt to changes in ecological conditions and maximise economic returns subject to ecosystem preservation constraints? In dryland grazing systems, for example, the resource can be most efficiently used by increasing stock rates in good conditions and destocking rapidly during droughts. Similar rules may be explored for other ecosystems - in fisheries, for example, by the use of particular harvesting strategies such as pulse fishing to optimally exploit the system.

Environmental Protection Measures

If benefit-cost analysis is properly applied to the design of development programs and projects and to resource management strategies, resource protection (insofar as it leads to the generation of economic benefits) will be undertaken up to the point where marginal protection benefits are equal to marginal protection costs (Hufschmidt et al., 1983).

The axiom that "prevention is cheaper than cure" is frequently valid in managing natural resources. Application of benefit-cost analysis can determine whether, in specific instances, this is necessarily the case.

If resource rehabilitation or protection is required, especially in the form of environmental improvement programs and projects, the timing of such programs and projects - within or at the end of the planning period - still has to be determined. Criteria governing the timing of resource protection measures within development projects also must be specified.

The problem of irreversible environmental impacts is an extreme case of resource degradation, in the sense that the restoration costs are prohibitive. If continued maintenance of existing stocks generates commensurate benefits, maintenance costs will be met, thus preventing extinction of natural resource stocks.

If the case is accepted for lowering the discount rate in natural resource management, on consistency grounds it should also be lowered for investments in natural resource protection and enhancement schemes. Long-term benefits of resource protection will thus receive due recognition in economic evaluations of options.

Concluding Remarks

As long as large deficiencies in knowledge and predictive capabilities prevail, and if social goals other than economic efficiency are to be attained, a conservative approach to sustainable development and natural resource management is recommended. Economic efficiency analysis can be applied within the constraints imposed by multiple objectives. Future options can be kept open by preserving natural capital stocks. This can be achieved by use of the policy and management guidelines described in this chapter.

References

Bishop, R. (1978), Endangered Species and Uncertainty: The Economics of a Safe Minimum Standard, American Journal of Agricultural Economics, vol. 60, no. 1.

Brooks, H. (1986), The Typology of Surprise in Technology, Institutions and Development, in: Clark, W.C. and Munn, R.E. (eds.), Sustainable Developments of the Biosphere, Cambridge University Press, Cambridge.

Ciriacy-Wantrup, S.V. (1952), Resources Conservation, Economics and Policies. University of California Press, Berkeley.

Clark, C. (1976), Mathematical Bioeconomics, Wiley, New York.

Crutzen, P., and Graedel, T. (1986), The Role of Atmospheric Chemistry in Environmental Development Interactions, in: Clark, W.C. and Munn, R.E. (eds.), Sustainable Development of the Biosphere, Cambridge University Press, Cambridge.

Dahlberg, K.W. (ed.) (1986), New Directions for Agriculture and Agricultural Research, Rowman and Allanheld.

Daly, H.E. (1973), Steady State Economics, Freeman, San Francisco.

Davis, B. (1987), The Practicalities of Policy Solutions, in: Chrisholm, A. and Dumsday, R. (eds.), Land Degradation: Problems and Policies, Cambridge University Press, Cambridge, pp. 335-340.

Dixon, J.D., Carpenter, R.A., Fallon, L.A., Sherman, P.B. and Manipoke, S. (1986), Economic Analysis of the Environmental Impacts of Development Projects, Asian Development Bank, Manila.

Dixon, J.A. and Hufschmidt, M.M. (1986), Economic Valuation Techniques for the Environment, John Hopkins University Press, Baltimore.

Dixon, J.A., James, D.E. and Sherman, P.A. (forthcoming), Economics of Dryland Management: Technical Guide and Case Study Reader.

Galeski, B., and Wilkening, E. (1987), Family Farming in Europe and America, Westview Press, London.

Gijsbers, D., and Nijkamp, P. (1988), Non-uniform Social Rates of Discount in Natural Resource Models: An Overview of Arguments and Consequences, Journal of Environmental Systems, vol. 17, no. 3, pp. 221-235.

Goodland, R., and Leduc, G. (1987), Neoclassical Economics and Principles of Sustainable Development, Ecological Modeling, vol. 38, pp. 19-46.

Groot, R. de (1987), Environmental functions as a Unifying Concept for Ecology and Economics, The Environmentalist, vol. 7, no. 2, pp. 105-109.

Holling, C.S. (1986), Resilience of Ecosystems: Local Surprise and Global Change, in: Clark, W.C. and Munn, R.E. (eds.), Sustainable Development of the Biosphere, Cambridge University Press, Cambridge.

Holling, C.S. (1987), Adaptive Environmental Assessment and Management, Wiley, New York.

Howe, C.W. (1979), Natural Resource Economics, Wiley, New York.

Hufschmidt, M.M., James, D.E., Meister, A.D., Bower, B.T. and Dixon, J.A. (1983), Environment, Natural Systems and Development: An Economic Valuation Guide, Johns Hopkins University Press, Baltimore.

IUCN (1980), World Conservation Strategy, IUCN, Geneva.

Jacobs, P. (1985), Achieving Sustainable Development. Landscape Planning 12, pp. 203-209.

James, D.E., Jansen, H.M.A. and Opschoor, J.B. (1987), Economic Approaches to Environmental Problems, Elsevier, Amsterdam/Brussels.

Kapp, K.W. (1974), Environmental Policies and Development Planning in Contemporary China and Other Essays, Mouton, The Hague/Paris.

Mandelbrot, B. (1977), The Analysis of Land Use Determinants in support of Sustainable Development, Collaborative Paper CD-88-1, IIASA, Laxenburg.

Mason, S.P., and Merton, R.C. (1984), The role of Contingent Claim analysis in Corporate Financing, Paper HBS 84-46, A.P. Sloan School of Management, MIT, Cambridge, MA.

NAS (1986), Acid Deposition: Long-Term Trends, National Academy Press, Washington.

Nijkamp, P., and Reggiani, A. (1988), Theory of Chaos in a Space-Time Perspective, Research Memorandum 1988-29, Faculty of Economics, Free University, Amsterdam.

Nijkamp, P. and Soeteman, F. (1988), Land Use, Economy and Ecology: Needs and Prospects for a Co-evolutionary Development, Futures, vol. 20, no. 6, pp. 621-635.

Norgaard, R.B. (1984), Co-evolutionary Development Potential. Land Economics, vol. 60, no. 2, pp. 160-173.

Opschoor, J.B. (1974), Economic Valuation of Environmental Pollution (in Dutch), Van Gorcum, Assen.

Opschoor, J.B. (1987), Sustainability and Change (Inaugural lecture, in Dutch), VU Boekhandel, Amsterdam.

Pearce, D.W. (1986), Foundations of an Ecological Economics, Int. Soc. Sc. Counc. Conference.

Pearce, D.W. (1988), Optimal Prices for Sustainable Development, in: Collard,D., Pearce, D. and Ulph, D. (eds.), Economics, Growth and Sustainable Environments, Macmillan, London.

Pearce, D.W., and Markanya, A. (1987), Marginal Opportunity Cost as a Planning Concept in Natural Resource Management, Annals of Regional Science, vol. XXI 3.
ibid, (1988), Sustainable Future: Natural Environments and the Social Rate of Discount. Project Appraisal, vol. 3 no. 3.

Randall, A., and Castle, E.N. (1985), Land Resources and Land Markets, in:

Kneese, A.V. and Sweeney, J.L. (eds.), Handbook of Natural Resource and Energy Economics, North-Holland Publ. Co., Amsterdam, pp. 571-620.

Repetto, R. (1986), World enough and Time, Yale University Press, New Haven/London.

Siebert, H. (1982), Nature as a Life Support System: Renewable Resources and Environmental Disruption, Journal of Ecos., vol. 42, no. 2, pp. 133-142.

Siebert, H. (1987), Environmental Economics, Springer, Berlin.

Solow, R.M. (1986), On the Intergenerational Allocation of Natural Resources. JEnv.Ecs.Man 4, pp. 1-24.

Walters, C. (1986), Adaptive Management of Renewable Resources, MacMillan, New York.

World Commission on Environment and Development (1987), Our Common Future, Oxford University Press, Oxford.

WRR (Scientific Council for Government Policy) (1988), Scope for Growth (in Dutch) Staatsuitgeverij, The Hague.

Chapter 4

TOWARDS SUSTAINABLE DEVELOPMENT: AN ENVIRONMENTAL PERSPECTIVE

R.E. Munn

Introduction

The relation between environment and development has been widely discussed in recent years. This has usually been in the Third World context, and often in conjunction with the terms **population** and **resources**.

Environment	Development
Population	Resources

Various connections and feedback loops exist amongst the elements in the boxes shown above, and it it clear that any discussion of development must take into account not only environment but also population and resources. Development could be limited, for example, by too few resources and too small or too large a population.

In this paper, the emphasis will be on sustainable development, which has been defined as 'development that meets the needs of the present without compromising the ability of future generations to meet their needs' (WCED, 1987). The sustainability question will be examined particularly from the ecological point of view. Furthermore, attention will be given not so much to the development problems of the Third World, which are so well described in the Brundtland Report (WCED, 1987), but to the re-development of industrialized countries, using Europe as an example. In this connection, the International Institute for Applied Systems (IIASA) in Austria has recently completed a sustainability study entitled: Future Environments for Europe: Some Implications of Alternative Development Paths (Stigliani et al., 1989) to which reference will be made later.

The main conclusion of the paper is that development is sustainable only if it is based on ecological principles and practices.

Sustainable Development: a Contradiction in Terms?

The phrase **sustainable development** has been criticized, for example by O'Riordan (1985), as a contradiction in terms. If development is equated with economic growth, this criticism is indeed justified: Malthusian limits prevent sustained growth in a finite world. In this connection, the argument has often been made that economic growth is a pre-requisite to meet the basic needs of the poor in 'developing countries': hence development is a 'global imperative' at least over the short term. Ultimately, however, uncontrolled economic growth will cause the quality of the environment to deteriorate, economic development to decline and the standard of living to drop. See Figure 1.

Of course, the word **development** does not necessarily imply growth. It may convey the idea that the world, society or the biosphere is becoming 'better' in some sense, perhaps producing more, or meeting more of the basic needs of the poor. The word therefore involves a value judgement. In principle, development could become sustainable through structural changes (economic, political, cultural or ecological) or a succession of technological break-throughs.

Two further points should be made. First, it is not realistic to believe that the entire world will ever achieve long-term sustainable development. Scientists can indicate the necessary (but not sufficient) preconditions that must be fulfilled if a country aspires to sustainability but there are certain to be individual countries devastated by plagues, wars and other disasters. A first principle for global sustainability is therefore that these catastrophes must not be allowed to reach continental or hemispheric dimensions, as might happen if there were a major nuclear war.

Secondly, the word **sustainable** must be carefully defined. A resource, which is sustainable when used or managed in certain ways may no longer be sustainable if these practices change (e.g., through drainage of wetlands, replacement of deciduous by coniferous forests, inflated prices of prime agricultural land near cities).

Having defined the use to which a system is to be put, the problem of determining quantitative criteria for maintaining sustainability in a steady-state world is relatively straightforward. Provided that concentrations of pollutants do not exceed environmental quality standards, for example, the resource of concern will continue to be renewable - all other factors remaining constant. Environmental toxicology is based on this principle.

Solutions to practical problems are of course complicated by the fact that the responses of members of a population of people or plants do not all

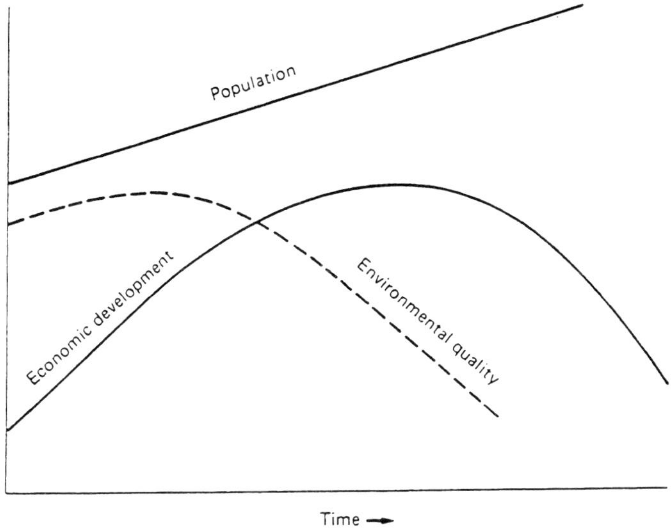

Figure 1: Schematic representation of the predicament of developing countries.

follow the same dose-response curves. In the case of ragweed pollen, for example, some people will not suffer from hay fever, no matter how high the concentrations. Also, the phrase 'all other factors remaining constant' rarely applies. In the case of trees and forests, for example, several types of environmental stresses may contribute to decline or death. These stresses have been classified functionally by Manion (1981), McLaughlin (1985) and others as **predisposing**, **inciting** and **contributing**. The general scheme shown in Table 1 (Manion, 1981; McLaughlin, 1985) illustrates that a great many stresses may contribute to the decline or death of trees. As an example, Manion (1981) cites the case of maple trees in eastern North America, which began to show signs of decline in the 1950s. The influencing factors were as follows:

Predisposing factors: Soil compaction, impeded drainage, road salt, air pollution.

Inciting factors: Drought, defoliating insects.

Contributing factors: Root rot, fungus

With so many stresses operating on a forest, how can one establish criteria for sustainability? One approach is through models but even if it were possible to construct a forest ecosystem model in which most of these stresses were included, it would be impossible to evaluate its performance because of the great natural variability in many of the environmental and ecological factors. In fact, variability is a characteristic feature of natural ecosystems, contributing to their resilience and sustainability.

	Types of influencing factors		
	Predisposing	Inciting	Contributing
Functional role	Chronic weakening	Triggering episodes	Accelerators
Stressing agents	Climate Soil moisture Genotype of host Soil nutrients Air pollutants Competition	Insect defoliation Frost Drought Salt Air pollutants Mechanical injury	Bark beetles Canker fungi Viruses Rood-decay fungi Competition

Table 1: Categories of factors influencing declines of forest trees (modified after Manion, 1981) (McLaughlin, 1985).

This is a general problem with most environmental/ecological processes.

Some Approaches to Achieving Ecological Sustainability

In the context of sustainable development, several approaches have been used to explore the ecological part of the problem. Some key phrases are:

Maximum sustainable yield (MSY) of a natural resource. The implication is that a renewable resource such as a fishery, forest or field crop may be safely harvested up to a certain level. As long as MSY is not exceeded, the resource is sustainable. This is the basis of international fishery agreements, for example.

Carrying capacity of a region. The implication is that a region can sustain a certain population of wild life, farm animals, or people. If the carrying capacity could be determined, it would be possible to set limits on the number of animals or people who should live in that region.

Assimilative capacity of a region. The implication is that it is acceptable to release pollutants into a watershed, air-shed or land-fill, provided that environmental standards are not violated. Several natural processes contribute to the self-purification of the environment, including transport, diffusion, chemical transformations (including radionuclide decay) and deposition. Pollution control agencies use this concept, even assigning transferable emission rights to industrial enterprises in some cases.

These approaches are sometimes useful for achieving short-term goals. However, difficulties must be mentioned in each case.

(1) Maximum sustainable yield

As emphasized by Walters (1986), the yield of a renewable resource increases steadily up to MSY but then drops rapidly, a situation illustrated schematically in Figure 2. An error in the calculation of MSY and/or a rare event such as an El Niño in the case of the Peruvian anchory fisheries could cause the resource to collapse.

(2) Carrying capacity

As above, there is the problem of rare events that might cause an ecosystem to exceed its carrying capacity and possibly trigger a major change. Odum (1983) therefore distinguishes between maximum carrying capacity and optimal carrying capacity, the latter being less vulnerable to environmental stresses (Brown et al., 1987). For human populations, the carrying capacity concept entails two additional difficulties. In the first place, human carrying capacity depends on the standard of living expected: a region at the subsistence level could sustain many more people than could a developed country. Secondly, carrying capacity depends on the energy subsidies provided, in the form of farm tractors, fertilizers, pesticides, transportation of produce to markets, cold storage, etc. Harwell and Hutchinson (1986) have discussed the hypothetical situation of sudden removal of these agricultural energy subsidies, all other factors remaining unchanged. The conclusion is that in most industrialized countries, there would be insufficient (and inexperienced) human labour to substitute for the tractor power currently available. In this connection, current labour inputs for grain production in developed countries average about 10 hours per hectare while in developing countries the inputs often average 1200 hours per hectare; the impacts of a removal of energy subsidies would not therefore be so great in developing countries.

(3) Assimilative capacity

Assimilative capacity varies greatly in both space and time. For example, the assimilative capacity of a river varies with flow rate and temperature. In the case of an air-shed, the pollution potential may differ by more than an order of magnitude from one day to another. In addition, many watersheds and air-sheds contain sub-compartments, e.g., living organisms or sediments, which accumulate and sometimes concentrate pollutants. Even if water, air or soil samples meet environmental standards, there may be flora or fauna that allow pollutants to enter food chains (lichens/ reindeer, zooplankton/commercial fish, etc.). In discussing this with

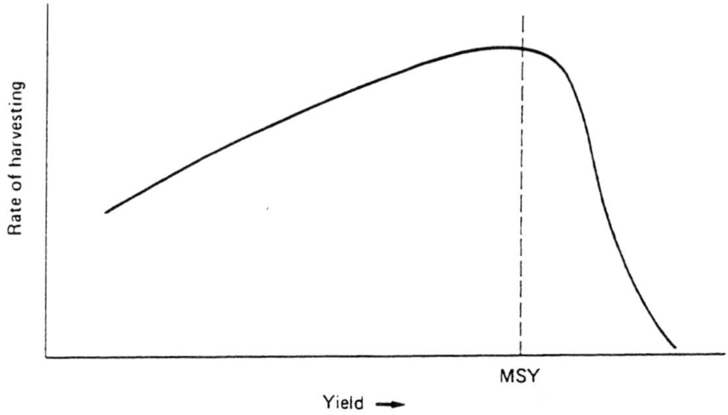

Figure 2: Schematic representation of the relation between rate of harvesting of a renewable resource and yield.

respect to the marine environment, Krom (1986) also questions the use of the word **assimilative**, preferring the phrase **accommodative capacity**.

A further point made by Regier and Grima (1984) is that 'a quantitative specification of an assimilative capacity is not a measure of a level of loading that does no harm, but rather a level of loading with which the harm done is unknown, or is ignored, or is deemed acceptable under the political status quo'. Regier (personal communication) in fact believes that the terms **carrying capacity** and **assimilative capacity** are political statements.

Resilience

There are various measures of the performance of an ecosystem or an engineering system. Some of the terms used include: complexity; stability; diversity; vulnerability; reliability; resilience; robustness; fragility; efficiency. The ecologist C.S. Holling was the first person to use the word, **resilience**, in an ecosystem context (Holling, 1973), arguing that the environment is never in a steady state. Trying to maintain the stability of an ecosystem over the long-term therefore imposes greater and greater stresses upon it if environmental conditions are changing. Holling introduced the term **resilience** as a measure of the ability of an ecosystem to adapt to a continuously changing environment, being able to absorb external shocks without major structural damage.

More recently, the resilience concept has been applied to socio-economic systems, and Burton et al. (1977) have suggested that primitive societies in the process of losing their traditional skills have little resilience. Highly advanced societies too have lost some of their resilience, being too dependent on technology.

Gradual change can be absorbed by most systems. However, fast change can be disastrous. In addition, many ecosystems contain non-linearities, which means that there are critical 'break points' beyond which, a small change in externalities will cause a large change in internalities. See Figure 2. A resilient system can accommodate both change and discontinuities.

Although resilience is a useful paradigm, it is difficult to express numerically. Holling (1986) suggests that patchiness is a characteristic of a resilient forest, although nobody has yet devised a quantitative method of determining whether one forest is more resilient than another. In the limit, of course, a monoculture has no resilience.

The net result of this discussion is that one of the characteristics of sustainable development ought to be its resilience.

Economic Realities

The phrase 'sustainable development' has an anthropogenic connotation. If there were no people on earth, development of the biosphere would only take place over geological time. In the present century, however, the sustainability of the biosphere is very much in the hands of society, which is changing the face of the Earth as never before.

Although sustainability is greatly to be desired, there are two types of factors that hinder the achievement of such a condition. In the first place, scientific knowledge and associated technology are not always sufficient, and there may be instances where irreversible trends are established through inadvertency; the application of environmental knowledge and models to extrapolate trends is always difficult. Secondly, there are economic, political and cultural realities that inhibit progress towards achieving sustainable development. A few examples can be given:

(1) Irrigation of semi-arid land leads to salinization of the soils. In order to increase crop yields, more and more irrigation is applied, and more and more land is damaged.

(2) As a fishery is over-harvested, the price of fish increases, encouraging fishermen to try even harder to catch fish.

(3) In 1980, 1283 million people cut down trees faster than trees grew, in order to get enough firewood (UNEP, 1987).

(4) Expanding urbanization consumes prime agricultural land, the life-support system that led to the creation of many cities.

These are only a few examples of the dilemmas facing society. Some rationalization has been attempted by arguing that economic development is a prerequisite for improvement in environmental quality, that society's first goal is to meet the basic needs of the poor. Sometimes the suggestion is also made that population growth is not necessarily bad.

The UNEP 1987 State of the World Environment report states these arguments very carefully (UNEP, 1987):

> Population, environment and development factors interact in different ways in different places. Not only the pace of development, but its content, location and the distribution of its benefits determine, in good measure, the state of the environment. These factors also influence the growth and distribution of population. Environmental resources provide the basis for development just as environmental factors constitute part of the improvement in the quality of life that development is meant to bring about. Similarly, the size of population, the rate of its growth and the pattern of its distribution influence the state of the environment, just as they condition the pace and composition of development.
>
> Population growth need not necessarily lower levels of living, impair the quality of life or cause environmental degradation. Global and historical assessments of the Earth's capacity and human ingenuity to produce goods and services have prompted some experts to project an optimistic outlook. Growth of world population has, in the past, been accompanied by a steady increase in the world's capacity to provide for the necessities and amenities of human life. People have to be fed, clothed and provided for, and this is achieved by people themselves. In that process, they use and develop the resources of the environment. Yet, just as some patterns of development have improved the human environment, others have tended to degrade it, at times irreversibly.

The Brundtland Commission (WCED, 1987) has much to say on sustainable economic development. The recommended strategies include:

- reviving growth;
- changing the quality of growth;
- meeting essential needs for jobs, food, energy, water and sanitation;
- ensuring a sustainable level of population;
- conserving and enhancing the resource base;
- reorienting technology and managing risk;
- merging environment and economics in decision making.

These strategies are not entirely consistent with strategies designed to promote ecologically sustainable development. For example, 'reviving growth' and 'conserving and enhancing the resource base' seem to involve some conflicts. As a matter of fact, there is a real need for more dialogue between economists and ecologists on these important questions, as indeed is recommended by the Brundtland Commission.

Human Disturbances of the Natural Environment

The spatial scales of human disturbances

Most environmental disturbances originate locally. In many cases, e.g., that of contaminated ground-water, the impacts have to be assessed and solutions have to be found locally, although similar problems may occur in many parts of the world. In other cases, the main impacts occur on the regional scale, as shown in the following diagram:

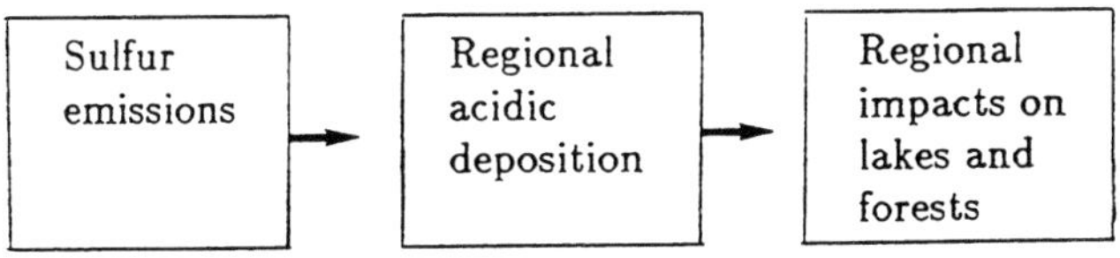

The atmospheric 'residence time' of sulfur is rather short, and the impacts are directly on the local and regional scale. Finally there are cases in which the issues are truly global. For example:

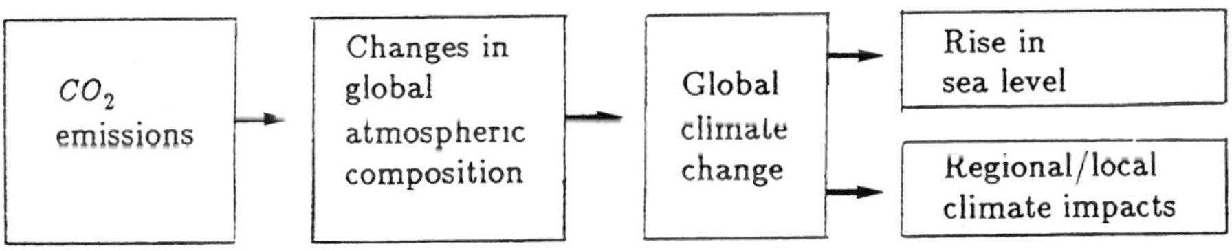

CO_2 has a relatively long atmospheric residence time and is diffused around the world, changing the global atmosphere and the global climate with resulting regional and local impacts.

Maurice Strong, former Executive Director of UNEP, coined the term **outer limits**. This was intended to suggest that although planet Earth has considerable resilience, we should be careful not to stress it too much too quickly.

Lovelock (1979) has elaborated an hypothesis termed Gaia, which assumes that the biosphere exerts homeostatic control of the chemical composition of the Earth's atmosphere. The process operates through the global biogeochemical cycles of nitrogen, oxygen, sulphur, phosphorous and other substances. This life support system contains many feedbacks which help to maintain global equilibrium. A recently discussed example of a homeostatic feedback involves dimethylsulphide (DMS), which is released by phytoplankton in the oceans (Charlson et al., 1987). If the climate were to warm, for whatever reason, the following changes would take place:

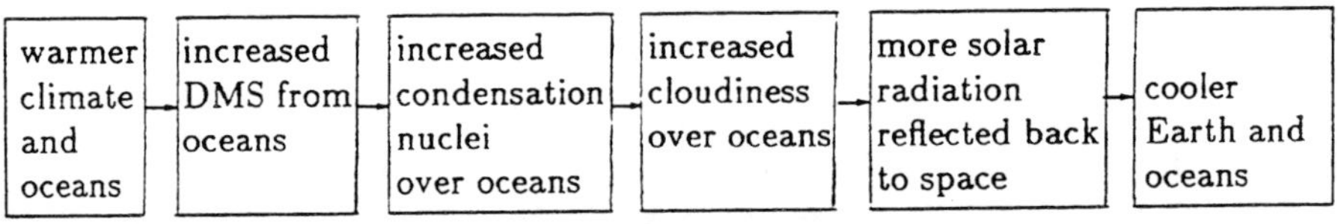

The fact that long-term global environmental equilibrium exists is illustrated by the annual average global temperature, which has varied by less than a degree from year to year over the last several centuries. McElroy (1986) notes that despite its natural variability, the Earth's environment has 'remained within narrow limits congenial to life for more than 4 billion years.'

Mankind is now changing the composition of the global atmoshpere very rapidly, and climatologists are in general agreement that the consequences will be enormous in the next hundred years. Major changes have already taken place in the stratospheric ozone layer, with a so-called ozone hole over Antarctica (Farman et al., 1985). There are also rapid increases in the concentrations of some of the atmospheric trace gasses: Figure 3 (Rasmussen and Khalil, 1984), for example, shows the rise in atmospheric methane concentrations that has occurred in recent decades in Antarctica.

Not everyone accepts the Gaia hypothesis but as emphasized by Holling (1986): "By being rooted in questions of regulation and stability through identifiable biological, chemical and physical processes, Gaia gives a direction for relevant scientific research - for disproof of the hypothesis if nothing else."

Responding to these problems, the International Council of Scientific Unions (ICSU) at its General Assembly in 1986 established a program IGBP (International Geosphere Biosphere Program), also called Global Change, which aims 'to describe and understand the interactive physical, chemical and biological processes that regulate the total Earth system, the unique environment that it provides for life, the changes that are occurring in the

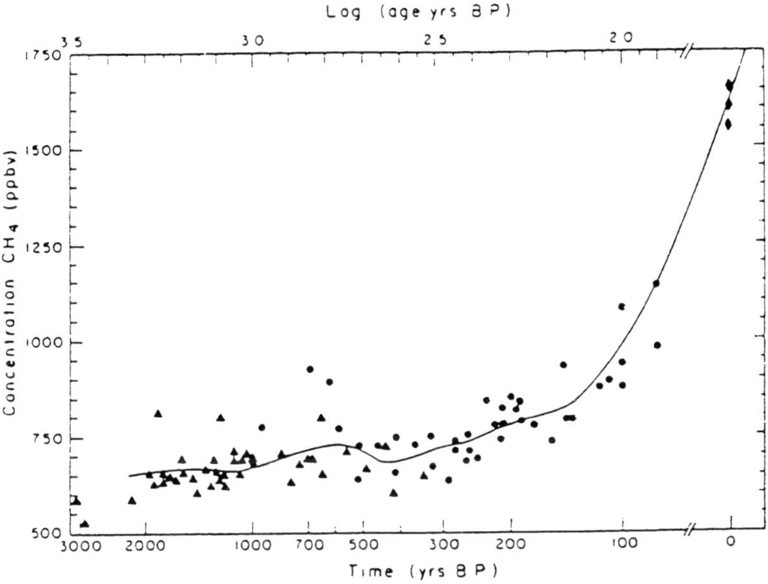

Figure 3: Methane concentrations obtained from ice cores in Greenland (circles) and Antarctica (triangles). The present concentrations of methane are shown on the right (diamonds) (Rasmussen and Khalil, 1984).

system, and the manner in which they are influenced by human actions' (ICSU, 1986). The IGBP program is strongly supported by many National Academies of Science and Royal Societies, and is expected to continue for at least two decades.

Regional problems

Human disturbances of the regional environment may arise as a result of:

(1) changes in the global environment, e.g., rising sea level;

(2) changes in the local environments within the region, causing cumulative regional impacts, e.g., acidic deposition, pollution of large rivers;

(3) a combination of (1) and (2), e.g., water quality of rivers in a warmer and drier climate.

Over the next 50 years, the type (3) situation will become more and more common, and is the most worrisome. Scientists are able to model the impacts of environmental stresses, one at a time, but they are not very good at predicting the integrated impacts of multiple stresses particularly where there are no historical analogies to guide them. Added to this is the fact that societies are losing resilience, not only primitive peoples, but also those that are highly developed: in the latter case, society's emphasis on increasing technological reliability reduces the frequency of 'accidents' but also reduces the preparedness of society to cope, if technology fails (Timmerman, 1981; Brooks, 1986).

Recalling that sustainable development must be ecologically sustainable, the first question to be asked in designing a development policy is 'Sustainable for what purpose?' So it is necessary to involve social scientists, determining the needs and wants of society and estimating how these may change in future decades. Frequently, a consensus will not arise but the differing viewpoints should be made transparent, so that alternative development pathways can be explored.

The differing viewpoints are partly due to a lack of knowledge of current conditions. Indicators of the economic health of an industry or of a country are generally available but there are few comparable figures on ecological 'health'. It has been suggested for example, that industrial productivity should be measured in terms of output per unit of resource used, rather than in terms of output per person per hour (UNEP, 1986). Elaboration of this simple idea into a practical system faces the difficulty that there is no generally accepted resource accounting system. We need some common measures of the rates at which an ecosystem is being degraded or restored.

An example of a regional sustainability study is given in one of the next sections.

Toward the Achievement of Sustainable Development

A summary of guiding principles

Some guiding principles for achieving sustainable development are as follows:

1. The outlook must be long-term - decades to centuries.
2. The approach must be integrative, with several issues treated at the same time. Incremental resource management should be discouraged.
3. The objective must be to make society and the biosphere more resilient to socioeconomic and environmental changes, keeping as many options open as possible.
4. The ideas of Ignacy Sachs on ecodevelopment ought to be more widely disseminated and practiced. Sachs makes the important point that a country should try to maximize the use of its available resources (human, non-renewable and renewable). Living within one's resources means that inequities between countries will continue to exist but

ecodevelopment ought to be at least a minimum goal for all nations.

Those concerned with serious day-to-day environmental problems are not easily convinced that they should worry about long-term global issues. There is indeed need for local action now, but this should be in larger context. In fact, the key slogans should be:

Think globally. Act locally!

Think long-term. Act now!

The construction of a power station, dam, highway or smelter, or the planting of a forest, is a commitment to the use (or mis-use) of natural resources for at least 50 years.

Some practical measures

The practical measures that could be taken towards achieving sustainable development are of three types:

- Measures designed to change the perception of society.
- Economic measures.
- Research strategies.

These subjects will be considered in turn.

Changes in the perceptions of society

By way of introduction, it will be useful to discuss a particular case - the forest industry in New Brunswick, Canada. Figure 4 (Regier and Baskerville, 1986) shows average diameters of tree stems of white pine and white spruce as a function of age in that province. Sawmill efficiency depends on stem size and when 300 year old white pine (point A in Figure 4) were harvested faster than recruited to that age, the industry shifted to smaller and smaller trees (down the upper curve) until it became economic to switch to white spruce (points B and C). However, continued over-exploitation is leading down the lower curve to point D, which represents current limits of economic acceptability.

Recognition of the essence of Figure 4 came in an interval of only five years (1975-1980) when senior New Brunswick officials in industry and government realized that a flow of quality trees could not be maintained at current harvesting rates. That relatively sudden change in perception was a prerequisite for serious consideration of a redevelopment strategy.

Perception of a problem is always a precondition for action, and in many cases, government, industry, business, the public and environmental scientists have differing perceptions of the seriousness of each one of an array of environmental issues on the political agenda at any particular moment. It is in fact necessary to involve the various 'players' in the

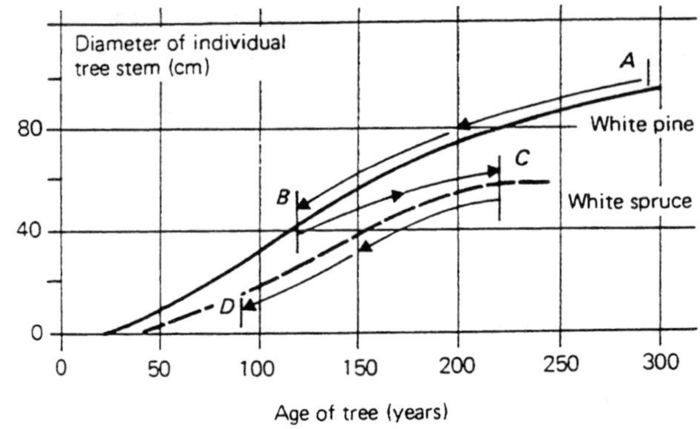

Figure 4: The average diameter of white pine and white spruce stems as a function of the age of the stems (Regier and Baskerville, 1986).

long-term environmental planning process, very much as Japanese manufacturing workers are involved in the operation of their factories. This is perhaps a radical idea but it is an essential one.

One of the obstacles to achieving consensus by various groups is of course, their differing perceptions of economic and social costs. In the case of the New Brunswick forest industry (Regier and Baskerville, 1986), one of the problems was that forest workers and the public believed at first that corrective measures could be taken at no cost, or at the expense of the companies concerned and they failed to recogize that the forest industry was losing its competitiveness in the world markets.

But how does one involve the major players mutually in a long-term environmental planning process? A promising approach is that of Brewer (1986), who uses the term **policy exercise** to describe a 'game' in which very senior management people are presented with dilemmas arising out of a series of ecological scenarios and not-impossible discontinuities. With the right mix of people and support staff, the policy exercise can:

- elaborate several policy options for handling the dilemmas, and identify the consequences of each;
- improve the quality of decision-making, with respect to political and policy consequences;
- identify important but poorly understood questions for further study.

Interesting (and possibly frightening) as the scenarios and discontinuities used in a policy exercise may be, they are not intended to be 'best estimates' of the future based on current knowledge. Instead, they are plausible or at least 'not impossible' vehicles for promoting dialogue amongst policy people with respect to highly complex issues. The future has been called the graveyard of scenarios, and Brewer suggests that 'a good scenario should be problem oriented, as determined by its success in focusing attention and opening up unknown possibilities. It should lay out biases for collective scrutiny and consideration.'

The 'policy exercise' approach is being tested at the International Institute for Applied Systems Analysis (Laxenburg, Austria).

Economic measures

At the outset, it should be emphasized that economic measures are tools for implementing an ecological strategy. Individual piecemeal actions designed to promote conservation, to control emissions of pollutants or to pay compensation to victims of environmental degradation may be important steps to take, but they will not in themselves ensure sustainability of the biosphere or of economic development.

There are, of course, many ways in which society can be induced or coerced into taking actions that promote the goal of sustainable development. Table 2 (Regier and Grima, 1984) provides a listing. In this connection, it should be emphasized that prohibition and regulation are not always the best approaches, particularly if the laws are difficult to enforce or if the fines imposed by the courts are insufficient. In fact, the best economic development strategy is an integrated approach in which all of the 13 measures listed in Table 2 are used. An example of an innovative approach to regional water management is the 1984 Treaty between Canada and the USA which ensures for a period of 80 years, that flooding of the Skagit Valley into British Columbia will not take place. The city of Seattle will not raise the height of the Ross Dam and in return, British Columbia will supply Seattle with electricity equivalent to that which would have been generated had the dam been raised. British Columbia will receive as payment for the electricity, the sums equivalent to the cost of construction, operation and maintenance of the dam (Press release October 29, 1984, Canadian Dept. of External Affairs).

Research strategies

We do not yet know how to ensure sustainable development of the biosphere, and there is a great challenge for ecologists and environmental scientists to develop research programs directed towards the central issues of sustainability, resilience and global change. The fundamental question to be asked is: What are the criteria for sustainable development of the biophere - for both natural and managed (e.g., agriculture, fisheries)

Table 2: Institutional or policy mechanisms, available to natural-resource and environmental mangers (Regier and Grima, 1984).

1. Prohibition, e.g., banning of a toxic substance.	7. Compulsory insurance to compensate for damages to victims of pollution.
2. Regulation, e.g., of commercial fishing intensity.	8. Effluent charges, e.g., a charge for waste disposal scaled according to the direct cost of treatment or to the indirect cost associated with deleterious impacts on a receiving ecosystem.
3. Direct government intervention to modify some ecosystem feature, e.g., liming of acidified lakes.	9. Resource rent, e.g, tax or charge on harvesters of a resource in order to recover a fair return for the owners (all the people) of the resource, and also to foster efficient use of the resource by discouraging overcapitalization.
4. Grants and tax incentives, e.g., a subsidy to industry for anti-pollution equipment, preferential parking for vehicles used in car pools.	10. Management of the demand, e.g, through rate structures involving marginal cost pricing and/or peak load pricing to improve overall efficiency of use and foster conservation.
5. Buy-back programs, e.g., government purchase of excess harvesting capacity in fisheries which is then retired in order to reduce effective fishing capacity.	11. Transferable development rights, e.g., limited rights to develop a particular area exchanged for broader rights to develop a different area as preferred by government.
6. Liability for compensation, e.g., losers of an amenity have the right to sue the despoilers of that amenity.	12. Transferable individual quotas, e.g., 'assimilative capacity' rights in the case of pollutants, harvest rights in the case of natural resources.
	13. International trade policies, e.g., tariffs, subsidies and foreign aid.

systems? Valuable work is reported on the rehabilitation of degraded ecosystems - but nearly always in the context of the present 'steady-state' world. Also, there is need for integration of issues, economic sectors and disciplines. One environmental framework that is being used successfully is the concept of biogeochemical cycling. The global cycles of nitrogen, sulphur, phosphorus and indeed water provide the essential life-support system for Earth. When these cycles are disrupted through human interventions, an interrelated series of problems arise, including acid rain, greenhouse-gas climate warming, stratospheric ozone depletion and land degradation by fertilizers. Much attention is currently directed towards study of the global biogeochemical cycles. See, for example, SCOPE 21 (1983) and SCOPE 29 (1986).

The IIASA Sustainability Study: Future Environments for Europe

The IIASA Environment Program has recently completed a study: Future Environments for Europe: Implications of Alternative Development Paths (Stigliani et al., 1989). The area studied was Europe, stretching from the Atlantic Ocean to the Ural mountains, and the time horizon extended to the year 2030. The goal was to promote policies that will contribute to sustainable development, both nationally and for Europe as a whole. The assumption was made that the future is very uncertain; the present must therefore be managed in such a way that technological, socio-economic and environmental developments/disruptions, some unexpected, will not produce major shocks on society, and in fact can sometimes be exploited to benefit at least part of the European population.

The study had four related components:

1. to develop plausible scenarios of Europe's socio-economic development, with special attention being given to a few low-probability, high-impact developments;
2. to characterize the resulting large-scale environmental/ecological transformations;
3. to assess the effectiveness of alternative steps that might be undertaken in efforts to manage these environmental transformations;
4. to identify the research and monitoring priorities for refining policy-relevant aspects of these environmental transformations.

Four alternative socioeconomic development pathways were selected for special study:

(1) **Present trends continuing in Europe and elsewhere**, i.e., slow economic

growth and modest success in slowing down environmental change, implying moderate climate warming by the year 2030.

(2) **High-growth economy in Europe and elsewhere**, with only lip service being paid to the environment, implying strong climate warming by the year 2030.

(3) **Environmentally friendly economy in Europe and elsewhere**, with only slight climate change by the year 2030.

(4) **Environmentally friendly economy in Europe but not elsewhere**, where there is high economic growth without adequate environmental protection, implying strong climate warming [as is pathway (2)].

The resulting large-scale environmental transformation were characterized in terms of 11 environmental policy dilemmas that could become major problems by the year 2030. These dilemmas were:

- Managing water resources in an era of climate change.
- Acidification of soils and lakes in Europe.
- Long-term forestry management and the possibility of a future shortfall in wood supply.
- Areas of Europe marginalized by mainstream economic and agricultural development.
- Sea level rise.
- Chemical pollution of coastal waters.
- Toxic materials buildup and the potential for chemical time bombs.
- Non-point-source emissions of potentially toxic substances.
- Transportation growth versus air quality.
- Decreasing multi-functionality of land owing to urban and suburban land development.
- Increasing summer demand for electricity, and the impact on air quality.

The major findings of the study are that:

(1) Continuation of present trends in economic development and environmental protection in Europe and elsewhere will not ensure sustainable development in Europe. (A high economic growth scenario in Europe and elsewhere would worsen this outlook.)

(2) Environmentally friendly development in Europe offers the hope of mitigating local and regional-scale problems such as acid rain specific to Europe. However, because of the linkages between the European and global environments, sustaining the European environment in the 21st

century cannot be achieved without sustaining the global environment. Thus Europeans in their own self-interest should provide leadership in creating environmentally friendly development pathways in all parts of the world.

Some of the resulting dilemmas facing European include:

- With warmer and drier summers, serious water shortages would occur, impacting on the industry, agriculture (irrigation) and energy (hydropower) sectors. From a global perspective, how should these disbenefits be shared? What, for example, would be the cost-effectiveness of building more reservoirs and how soon should they be built?
- What practical policies ought to be adopted today to protect the European boreal forests from being destroyed shortly after year 2030 assuming (1) continuation of the current stresses of acid deposition and ozone; (2) to additional major stresses - due to climate warming and the exhaustion of the aluminum buffering capacity of the soils?
- What practical land-use policies ought to be adopted today assuming that by the year 2030, a significant fraction of the agricultural land in Europe will have lost its fertility (due to soil erosion, salinization, etc.) or will have been swallowed up by cities, and that crop yields from the remaining lands will have diminished due to a reduction in soil moisture and an increase in the frequency of ozone episodes?
- Public and media perception, whether right or wrong, is rather suddenly focussed on ecological degradation, which is generally agreed to be unaccpetable despite the massive sums of money being spent on pollution control. What earlier steps should have been taken to avoid this loss of credibility, and what should policy people do now?

These are examples of the policy questions being posed to very senior people in government and industry sitting together from time to time with European ecologists at IIASA.

Some Final Remarks

A necessary (but not sufficient) condition for sustainable development is that it be ecologically sustainable. However, criteria for global, regional or even single-ecosystems sustainability are difficult to formulate.

The recent literature on sustainable industrial development goes part way towards the goal of ecological sustainability. See, for example, CONCAWE (1987), WCED (1987) and WICEM (1984), in which the following points are emphasized:

- Environmental impact assessment should be required prior to approval being given to industrial development proposals.
- Good environmental management practices should be followed at all industrial sites.
- Better and cheaper emission-control technologies should be developed (The expressions **best available means** and **best practicable means** are widely used, suggesting that there are technological and economic limits to the efforts that industry can make towards sustainable development.)
- Environmental assessments should be broadened to include 'policies and programs, especially major macroeconomic, finance and sectoral policies that induce significant impacts on the environment' (WCED, 1987).
- The global commons should be preserved; for example, industrial releases of chlorofluorocarbons should be reduced.

These measures are helpful and can be justified on economic grounds. In CONCAWE (1987), for example, it is stated that in the European Community, the pollution control sector produces well over 1% of Community GNP, and that the national environmental policy of the FRG has resulted in the creation of 440,000 jobs. The Executive Director of UNEP states: "It is evident that economic growth and environmental protection go happily hand in hand' (Tolba, 1987). Nevertheless it should be emphasized that these measures in themselves will not ensure sustainability of the biosphere. One reason for this is that many individually small but cumulating impacts (CEARC/NRC, 1985) may lead to long-term environmental degradation in the quality of life. For example, the construction of a thermal power station may meet environmental standards but this development may trigger urban and industrial growth in the surrounding area; although these changes may be widely perceived to be 'good', a process may have been set in motion that will some day cause the carrying capacity of the region to be exceeded. A second and more important problem is that development/redevelopment planning usually assumes that the externalities remain unchanged. The steps taken to achieve sustainable development may therefore not be particularly effective in addressing the most important issues.

Because new technology diffuses very quickly around the world, and because environmental side effects may sometimes not be recognized for 10-20 years, it is important that international environmental assessments of new technologies be undertaken on a continuing basis. The recent Swedish initiative SIIESTA is welcomed (SIIESTA, 1988).

Finally, dialogue should be encouraged between those who seek sustainable **economic** development and those who strive towards sustainable **ecological** development. In this connection, there is need for both groups to formulate quantitative criteria for sustainable development; in the case of ecological sustainability, this need has been particularly emphasized by Soviet scientists such as Yu. Izrael (Izrael et al., 1988).

References

Brewer, G.D., (1986), Methods for policy exercises, in: Clark, W.C. and Munn, R.E. (eds.), Sustainable Development of the Biosphere, Cambridge Univ. Press, Cambridge, UK, pp. 445-473.

Broecker, W.S., Unpleasant surprises in the greenhouse? Nature 123-126, 1987.

Brooks, H., (1986), The typology of surprises in technology, institutions and development, in: W.C. Clark and R.E. Munn (eds.), Sustainable Development of the Biosphere, Cambridge University Press, Cambridge, UK, pp.

Brown, B.J., Hanson, M.E., Liverman, D.M. and Merideth, R.W. (1987), Global sustainability: towards definition, Env. Manag. 11:713-719.

Burton , I., Kates, R. and White, G. (1977), The Environment as Hazard, Oxford Univ. Press, Oxford, UK.

CEARC/NRC (1985), Proc. Workshop on Cumulative Environmental Effects: a Binational Perspective, Canadian Env. Assess. Res. Council, Ottawa, 175 pp.

Charlson, R.J., Lovelock, J.E., Andreae, M.O., and Warren, S.G. (1987), Oceanic phytoplankton, atmospheric sulphur, cloud albedo and climate, Nature: 326:655-661.

CONCAWE (1987), Environmental Protection in the 1990s: Main Speaker Presentations, Rep. No. 9/87, CONCAWE, The Netherlands, 60 pp.

Farman, J.C., Gardiner, B.G. and Shanklin, J.D. (1985), Large losses of total ozone in Antarctica reveal seasonal ClO_x/NO_x interactions, Nature 315:205-210.

Harwell, M.A. and Hutchinson, T.C. (1986), Environmental Consequences of Nuclear War, Vol. II, SCOPE 28, John Wiley, Chichester, UK, 523 pp.

Holling, C.S. (1986), The resilience of terrestrial ecosystem: local surprise and global change, in: Clark, W.C. and Munn, R.E. (eds.), Sustainable Development of the Biosphere, Cambridge Univ. Press, Cambridge, UK, pp. 293-317.

ICSU (1986), Resolution adopted at General Assembly in Berne, and subsequent press release on IGBP, ICSU, Paris, France.

Izrael, Yu.A., Antonovsky, M. and Semenov, S.M. (1988), On man-induced instability of the biosphere, Proc. of Conf., Ecologically Sustainable Development of the Biosphere, Goskomgidromet, Moscow (in Russian).

Krom, M.D. (1986), An evaluation of the concept of assimilative capacity as applied to marine waters, Ambio, 15:208-214.

Lovelock, J.E. (1979), GAIA: A New Look at Life on Earth, Oxford University Press, Oxford, UK.

Manion, P.D. (1981), Three Disease Concepts, Prentice Hall Inc., Englewood Cliffs, N.J.

McElroy, M.B. (1986), Change in the natural environment of the Earth: the historical record, in: Clark, W.C. and Munn, R.E. (eds.), Sustainable Development of the Biosphere, Cambridge Univ. Press, Cambridge, UK, pp. 199-212.

McLaughlin, S.B. (1985), Effects of air pollution on forests, J. Air Poll. Cont. Assoc. 35:512-534.

Odum, E.P. (1983), Basic Ecology, Saunders College Pub., New York.

O'Riordan, T. (1984), Future directions for environmental policy, IIUG dp 84-14, Int. Institute for Env. and Society, Berlin.

Rasmussen, R.A. and Khalil, M.A.K. (1984), Atmospheric methane in the recent and ancient atmospheres: concentrations, trends and interhemispheric gradient, Geophys. Res. 89:11599-11605.

Regier, H.A. and Grima, A.P., The nature of Great Lakes ecosystems, Int. Bus. Lawyer, June 1984 issue, pp. 261-269.

Regier, H.A. and Baskerville, G.L. (1986), Sustainable redevelopment of regional ecosystems degraded by exploitive development, in: Clark, W.C. and Munn, R.E. (eds.), Sustainable Development of the Biosphere, Cambridge Univ. Press, Cambridge, UK, pp. 75-101.

SCOPE 21 (1983), The Major Biogeochemical Cycles and their Interactions, Bolin, B. and Cook, R. (eds.), John Wiley, Chichester, UK, 554 pp.

SCOPE 29 (1986), The Greenhouse Effect, Climate Change and Ecosystems, Bolin, B., Doos, B.R., Jager, J. and Warrick, R.A.(eds.), John Wiley, Chichester, UK, 574 pp.

Stigliani, W.M., Brouwer, F.M., Munn, R.E., Shaw, R.W. and Antonofsky, M. (1989), Future environments for Europe: some implications of alternative development paths, Sci. Total Env. (in press), 102 pp. (See also IIASA Executive Report 15, Feb 1989, 17 pp.)

SIIESTA (1988), An International Institute for the assessment of environmental sound technologies, Ministry of the Environment and Energy, Stockholm, Sweden, 470 pp.

Timmerman, P. (1981), Vulnerability, Resilience and the Collapse of Society, IES Env. Mon. No. 1 (Institute for Environmental Studies, University of Toronto, Toronto, Canada).

Tolba, M. (1987), Introductory statement of the Executive Director to the Governing Council, UNEP GC.14/4/ Add.8,8 pp.

UNEP (1986), Output per unit of resources is what counts - Tolba, UNEP News, Sept - Oct issue, UNEP. Nairobi, Kenya, p. 6.

UNEP (1987), The State of the World Environment, UNEP, Nairobi, Kenya, 76 pp.

Walters, C. (1986), Adaptive Management of Renewable Resources, MacMillan, London, UK, 374 pp.

WCED (1987), Our Common Future, Oxford University Press, Oxford, UK, 383 pp.

WICEM (1984), World Industry Conference on Environmental Management, Ind. and Env. 5 39 pp.

Chapter 5

STEADY-STATE AND GROWTH CONCEPTS FOR THE NEXT CENTURY[1]

H.E. Daly

Introduction

Three conceptual issues seem to me critical for clear thinking about economic development and the environment in the next century. I will begin by stating the three issues briefly, and then discuss each one in turn, along with the relations among them.

(1) The first issue is whether the basic conceptual starting point of economic analysis is to be the circular flow of exchange value, as it presently is, or the one-way entropic throughput of matter-energy. The latter concept is virtually absent from economics today, yet without it it is impossible to relate the economy to the environment. It is as if biology tried to understand animals only in terms of their circulatory system, with no recognition of the fact that they also have digestive tracts. The metabolic flow is not circular. The digestive tract firmly ties the animal to its environment at both ends. Without digestive tracts organisms would be self-contained perpetual motion machines. Likewise for an economy without an entropic throughput.

(2) The concept of optimal allocation among alternative uses of the total resource flow (throughput) must be clearly distinguished from the concept of an optimal scale of total resource flow relative to the environment. Under ideal conditions the market can find an optimal allocation in the sense of Pareto. But the market cannot find an optimal scale any more than it can find an optimal distribution. The latter requires the addition of ethical criteria; the former requires the addition of ecological criteria. The independence of allocation from distribution is widely recognized; the independence of allocation from scale is not as widely recognized, but is easily understood. In theory we can double the population and the per capita resource use rate, or cut them in half, and the market will still grind out a Pareto optimal allocation for every scale. Yet the scale of the economy is

---

[1] The views expressed in this paper are those of the author and should not in any way be attributed to the World Bank or any other institution. I am indebted to S. Davis, R. Goodland and E. Wessels for helpful comments.

certainly not a matter of indifference. A boat that tries to carry too much weight will still sink even if that weight is optimally allocated. Allocation is one thing, scale is something else. We must deal with both, lest even the efficiently allocated weight of the economy sink the environment. Economics has tried to reduce scale issues to matters of allocation (just get the prices right), and has thereby greatly obscured the relation between the economy and the environment.

(3) While an optimal allocation can result from the individualistic marketplace, the attainment of an optimal scale will require collective action by the community. In today's world community exists mainly at the national and subnational levels. The sense of national community has atrophied under the ideology of individualistic cosmopolitan free trade. Protection of national community has been lost in the legitimate fight against the economic protectionism of special interest groups. The scale of economic interdependence has become global, while the sclae of community control has remained national and subnational. This lack of congruence permits great scope for environmental damage resulting from individual decisions that remain unchecked by community control. Economists' complacency on this score rests on the misapplication of the principle of comparative advantage to a world in which the major premise of that principle, namely capital immobility between nations, no longer holds. Whether capital was in fact all that immobile in Ricardo's time can certainly be doubted, but what is beyond doubt is the logical necessity of that assumption for the doctrine of comparative advantage.

Underlying these three issues is the theme of steady state versus growth as the normal, healthy condition of an economy. Our growth-bound way of thinking makes it hard for us to admit the concept of throughput, because it brings with it the first and second laws of thermodynamics, which have implications that are unfriendly to the continuous growth ideology. The circular flow raises no such problems. The growth ideology is extremely attractive politically because it offers a solution to poverty without requiring the moral discipline of sharing. Also the obvious implication of recognizing an optimal scale is that growth should stop once the optimum is reached - that growth beyond the optimum scale is "anti-economic growth". Optimal allocation has no such growth-limiting implications. Likewise it comes as an unwelcome insight to recognize that growth fostered by the global integration of the world economy may impose an unacceptable cost in terms of sacrifice of the economic basis of national community and the ability of that community independency to set its own environmental and social standards. Comparative advantage is pressed into service to substitute comfortable logic for a painful reexamination of the facts presupposed by that logic.

The three issues I am raising are not difficult, arcane, or esoteric- they are no more than common sense - but it is hard for us to think clearly about them because doing so threatens the absolute priority of growth as the North Star of economic policy. Although the three issues are separable they are also related. Once throughput is recognized as a fundamental and indispensible concept, then the question of its optimal scale within a finite ecosystem naturally arises, along with the recognition that the question is different from that of optimal allocation. Once we face up to the question of limiting scale we recognize the collective or social nature of the task and the futility of leaving it up to the individualism of the market which can only deal with allocation, and which, under the flag of free trade is weakening the bonds of community within nations. Thus the very possibility of community control as a supplement to market individualism is being undermined. It remains to spell this out in more detail, while elaborating each of the three issues.

Much confusion could be avoided if we would agree to use the word "growth" to refer only to the quantitative scale of the physical dimensions of the economy. Qualitative improvement could be labeled "development". Then we could speak of a steady-state economy as one which develops without growing, just as the planet earth, of which the economy is an open subsystem, develops without growing. Growth of the economic organism means larger jaws and a bigger digestive tract. Development means more complete digestion and wiser purposes. Limits to growth do not imply limits to development.

The word "community" also needs clarification. Community exists at many levels from family to neighbourhood, to city, state, nation and world. Each larger unit is thought of as a federation of the smaller units, as a "community of communities". The focus here on the national level is not meant to imply that community is strongest at that level, or that world community is not a worthy goal. It simply reflects the fact that the nation is the main unit of authority for making economic policy, especially those policies that will either strengthen or weaken community at all levels in the future.

Entropic Throughput of Matter-Energy Versus the Circular Flow of Exchange Value

Georgescu-Roegen (1971) has pointed to "the standard textbook representation of the economic process by a circular diagram, a pendulum movement between production and consumption within a completely closed system" as proof of the mechanistic epistemology of modern economics. There is only reversible motion, a circular flow, and no recognition of irreversible entropic change. There is only mechanical time, no historical time. This does not mean that economists deny historical time or the entropy law - but it does mean that they cannot deal with them at the most basic conceptual level of economics, and have to introduce them in ad hoc and unintegrated ways outside the structure of formal models.

In addition to the mechanistic epistemology there was a practical reason for ignoring the entropic throughput. Economists are interested in scarcity, and during the formative years of economic theory the environment was considered, with some reason, an infinite source of raw materials and an infinite sink for waste materials, so the throughput was not scarce and was naturally abstracted from. Only scarce items entered into exchange, and exchange value flowed in a circle. So the circular flow became the paradigm within which we sought to understand the economic process. Once the throughput itself became scarce then the circular flow vision became **economically**, as well as physically, misleading. It totally obscured the emerging scarcity of environmental services. The circular flow has no beginning and no end, no points of contact with anything outside itself. Therefore it cannot possibly register the costs of depletion and pollution, nor the irreversible historical effects induced by the entropic nature of the throughput.

The concept of throughput was introduced into economics by Kenneth Boulding (1966), and more fully elaborated and integrated into economic theory by Georgescu-Roegen (op. cit.), who called it the "metabolic flow" and emphasized the manifold consequences of its entropic nature. Others (Kneese, Ayres and d'Arge, 1970) have paid respect to the concept by way of emphasizing the importance of "material balances", thus recognizing the constraint on the economic process of the first law of thermodynamics, but neglecting that of the second law. The first law is consistent with the circular flow vision - the same indestructible building blocks of matter-energy could simply cycle faster and faster around the production-consumption loop. Nothing gets used up. But the second law says that something does get used up - not matter-energy itself, but its capacity for rearrangement. Energy is conserved, but its capacity to do work is used up. To my knowledge no economics textbook has paid attention to any of these important contributions. Instead they continue to perpetuate the circular flow vision without so much as a reference to the concept of throughput (see Heilbroner and Thurow, 1981). Naturally if the very concept of throughput is not admitted it will be impossible to consider the issue of its optimal scale, a theme to which we now turn (see Daly, 1985).

Optimal Allocation Versus Optimal Scale

Standard economics is about the optimal allocation of resources, which in this broad sense includes labor and capital as well as natural resources. But natural resources are not viewed as the components of an entropic metabolic flow from and back to the environment. Rather they are seen as building blocks which are indestructible elements in the circular flow. Allocation of these elements among competing uses is the only question raised for standard economics by its partial recognition of throughput. As mentioned earlier a Pareto optimal allocation can be achieved for any scale of population and per capita resource use. The concept of economic efficiency is indifferent to the scale of the economy's physical dimensions, just as it is indifferent to the distribution of income. Equity of income distribution and sustainability of scale are outside the concept of market efficiency. Yet the environment is sensitive to the physical scale of the economy, and human welfare is sensitive to how well the environment functions. To optimally allocate resources at a non-optimal scale is simply to make the best of a bad situation. If the economy continues to grow it is simply to keep making the best of an ever worsening situation. This anomaly is absent from the circular flow vision: if the economy is an isolated system with no dependence on its environment then naturally it can never exceed the capacity of the environment. Its scale relative to the environment is a matter of complete indifference. But once we recognize the central importance of the throughput, then we must concern ourselves with its optimal scale as well as its optimal allocation.

Optimal scale of a single activity is not a strange concept to economists. Indeed microeconomics is about little else. An activity is identified, be it producing shoes or consuming ice cream. A cost function and a benefit function for the activity in question are defined. Good reasons are given for believing that marginal costs increase and marginal benefits decline. The message of microeconomics is to expand the scale of the activity in question up to the point where marginal costs equal marginal benefits, a condition which defines the optimum scale. All of microeconomics is an extended variation on this theme.

When we move to macroeconomics, however, we never again hear about optimal scale. There is no optimal scale for the macro economy. There are no cost and benefit functions defined for growth in scale of the system as a whole. It just doesn't matter how many people there are, or how much they each consume. If every micro activity has an optimal scale then why does not the aggregate of all micro activities have an optimal scale? If I am told in reply that the reason is that the constraint on any one activity is the fixity of all the others and that when all economic activities increase proportionally the restraints cancel out, then I will invite the economist to increase the scale of the solar flux and the hydrologic cycle in proportion to agriculture and industry. I will admit that if the ecosystem can grow indefinitely then so can the aggregate economy. But until the diameter of the earth begins to grow at a rate equal to the rate of interest one should not take this answer too seriously. The total absence in macroeconomics of the most basic concept of microeconomics is a glittering anomaly, and it is not resolved by appeals to the fallacy of composition. What is true of a part is not **necessarily** true for the whole, but it can be and usually is unless there is some aggregate identify or self-cancelling feedback at work. (As in the classic examples of all spectators standing on tiptoe to get a better view and each cancelling out the better view of the other, or in the observation that while any single country's exports can be greater than its imports, nevertheless the aggregate of all exports cannot be different than the aggregate of all imports). But what analogous feedback or identity is there that allows every economic activity to have an optimal scale while the aggregate economy remains indifferent to scale?

In the circular flow vision there is an aggregate identity - total expenditures equal total receipts, one person's expense is another person's income. Costs and benefits are conflated in transactions. In circular flow accounting we add up transactions rather than compare costs and benefits at the margin, so the question of an optimal scale of the circular flow never arises. It is the throughput that has an optimal scale. When growth exceeds the optimal scale we experience generalized pervasive externalities, such as the greenhouse effect and acid rain, which are not correctible by internalization of localized external costs into a specific price. As growth increasingly turns previously free goods into scarce goods the standard solution is to put positive prices on the newly scarce goods. Once a good has become scarce it is important that it have a positive price in order to be properly allocated. But there is a prior question: How do we know that we were not better off at the previous scale when the good was free and its proper price was zero? In both instances the prices were right. But that does not mean that the scale was right. Furthermore, the new exchange value created when previously free goods become scarce reflects a cost, not a benefit as currently reckoned. The classical economist Lauderdale recognized that private riches could expand while public wealth declined. This perversity will occur whenever formerly abundant objects with great use value, but no exchange value, become scarce and thus acquire exchange value. Although scarcity is necessary for value in the sense of measurable exchange value, "the common sense of mankind would revolt at a proposal for augmenting wealth by creating a scarcity of any good generally useful and necessary to man". (See Lauderdale, 1819). The revolt has been slow in coming, but let us hope that Lauderdale was right!

Optimal allocation at least has a definition, however restrictive and limited in relevance it may be. But how do we define optimal scale? This is an enormous question that involves not only much greater knowledge of carrying capacity and ecological relations, but also much clarification and deeper understanding of our own purposes. Many economists keep the scale question out of sight by rejecting the concept of carrying capacity on the grounds that it is not clearly defined. But by that criterion they should also refuse to talk about "time", one of the most difficult to define concepts of all. Some say it is absolute, others say it is relative, still others insist that it is pure illusion. Even "money" should not be spoken of, since what is **really** money, M1 or M2? Or M1A? One of the temptations of debate is to demand an unreasonable standard of precision for concepts that have troublesome implications for one's position, while being more informal and relaxed in the company of concepts known not to raise impolite questions.

But there is one thing we know about the optimum scale: it must at least be sustainable. So for the time being we can devote our practical policies toward sustainability, while we puzzle over the deeper philosophical issues of optimal scale.

It is widely believed by persons of diverse religious convictions that there is something fundamentally wrong in treating the earth as if it were a business in liquidation. The value of sustainability is so basic that it is usually tacitly assumed in our thinking. To most of us it comes as a surprise to learn that sustainability is built into the very concept of income. J.R. Hicks (1948) defined income as the maximum amount that a person or a nation could consume over some time period and still be as well off at the end of the period as at the beginning. Hicks further argued that the whole practical reason for calculating income is to have a guide as to how much we can consume year after year without eventually impoverishing ourselves. Income equals maximum sustainable consumption.

Until such time as we can develop national measures or at least perceptions of the costs and benefits of growth that can be compared at the margin to define an optimal scale, we will have to content ourselves with the more modest goal of making NNP a better approximation of Hicksian or "sustainable income". The latter term should logically be avoided as a pleonasm, but the redundancy is justified because we have forgotten the proper meaning of income. To be specific, we could not consume NNP year after year without impoverishing ourselves. There are two main reasons. First, NNP treats consumption of depletable natural capital no differently form other income. Consumption of man-made capital is not treated as income because it is not sustainable. Consumption of natural capital is not sustainable either, but nevertheless is treated as income. Second, defensive expenditures incurred to protect ourselves from the unwanted side effects of other production are counted as final expenditures and hence income, when in reality they are intermediate expenditures (costs of producing the commodity whose production generated the unwanted side effects). If, in addition to true income, we consume every year an additional amount equal to the sum of defensive expenditures and natural capital depreciation, then we will impoverish ourselves. The whole practical reason for calculating income in the first place is contradicted by present practice. Subtracting an estimate of these two unsustainable categories from NNP would give us a much better approximation of Hicksian income.

National income accountants are understandably peeved by instant experts who want to redo the national accounts to better reflect their pet cause. I want to emphasize that this proposal does not in the least try to convert the NNP into a welfare measure. Rather it is the call of some national income accountants to their fraternity to take more seriously their own first principles, to do a better job of measuring income in the interest of avoiding impoverishment by excessive consumption. Measuring sustainable consumption, or Hicksian income, correctly is at least a step toward defining optimal scale.

National Community Versus Free Trade

In a purely individualistic world there would be no reason to calculate national income. So the fact that we worry about national income at all is already some evidence against the view of atomistic individualism. Community sometimes seems to dissolve into nothing more than the aggregate of individuals. But the individual also dissolves into a node of intersecting social relationships. These relationships are internal to the very identity of the individual, and do not merely externally relate one independently defined individual to another. John, son of Henry and Martha, brother to Samuel, husband to Susan, friend of William and David, Republican, Presbyterian, employee of Exxon, citizen of Australia, etc. What is "John" independent of these relations? There is something that still exists, but it is a bit like the hole in the donut, a center, but a rather empty one. There can be no community without individuals, and there can be no individual identity without the context of community that provides the network of relations by which individuality is defined. The concept of "person-in-community" captures the duality that is denied by both individualism and collectivism. Some things ought to be done individually, others collectively. Among the things which individual action can attain, given a minimal social context, is an optimal allocation of resources via the price system. Among the things requiring collective action is that minimal social context itself, which is no longer quite so minimal as before since it now includes the choice of scale. At the very time when extra demands are being placed on our capacity for community action the basis of national community is being eroded in the name of free trade, which temporarily postpones the scale issue by in effect erasing national boundaries and allowing nations to draw on the world ecosystem rather than on just that part of it within their national boundaries.

Growth, International Trade, and Destruction of Community. No economic tenet commands greater consensus among economists than the doctrine of free trade and comparative advantage (see Cobb and Daly, 1989, and Culbertson, 1984). Academic economists have become so enamored of the logical argument for comparative advantage (a beloved **pons asinorum** in the standard pedagogy), that they have failed to notice that the fundamental assumption of the whole argument, capital immobility between nations, no longer represents even a weak approximation to the real world. Even though nine out of ten arguments for tariffs, quotas, or other trade restrictions can be exposed as special pleading, there is nevertheless a compelling case against free trade which Adam Smith and David Ricardo themselves would have found convincing had they lived in a world of free capital mobility, rapid population growth, and pervasive externalities resulting from the large scale of throughput.

Free trade remains dogma because it serves the Central Dogma of growth. But it does not serve community, and in fact works strongly toward the destruction of community. To see this, consider, for example, what US capitalists are in effect saying to US laborers. They are saying that US labor must complete in the world labor market with the masses in the third world, otherwise capital will have to move abroad. Nowadays technology and managerial skills move with capital. Capital mobility means that wages will tend to equality worldwide in the interest of efficiency and growth. Wages in the US will decline. Wages in the third world will rise - or will they? Not really, because the supply of labour is very large and becoming larger thanks to the demographic explosion. Also half of the world's supply of low-wage labours is in India and China, countries which have not traditionally competed in the world market, but are now entering. The isolationist policies of Chairman Mao and the restrictive trade policies of British imperialism are over. The equilibrium wage under free trade will be the third world level. Capital will benefit from cheap labor at home as well as abroad. Low wages mean a low standard of living for the great majority of our citizens.

We are assured by mainstream economists that this is just an irrational phobia of poor ignoramuses who have never been able to understand comparative advantage. But it is **they** who do not understand comparative advantage, because they are trying to apply a doctrine that depends on capital immobility between nations to a world in which capital crosses national boundaries at the speed of light in response to a tenth of a percent difference in rate of return! Many international transactions are now internal transfers between divisions of multinational corporations and take place with even greater ease.

Ricardo - in his **Political Economy of Taxation** (p. 136) - points out that if capital were as freely mobile between England and Portugal as between London and Yorkshire, then trade between the two countries would be governed by the labor theory of value (absolute advantage in terms of labor costs) rather than comparative advantage. Everything that differentiates domestic from international trade depends for Ricardo explicitly on the international immobility of capital (labor immobility between nations was taken for granted). Moreover, Ricardo's explanation of capital immobility invokes the theme of community:

"Experience, however, shews, that the fancied or real insecurity of capital, when not under the immediate control of its owner, together with the natural disinclination which every man has to quit the country of his birth and connexions, and intrust himself with all his habits fixed, to a strange government and new laws, check the emigration of capital. These feelings, which I should be sorry to see weakened, induce most men of property to be satisfied with a low rate of profits in their own country, rather than seek a more advantageous employment for their wealth in foreign nations." (op. cit. pp. 136-137).

For Ricardo it is the force of community that keeps capital at home even in the face of higher profits abroad. Furthermore he affirms that he would be sorry to see these feelings of community weakened. Perhaps he already suspected that they would be weakened by the individualistic postulates of classical economics, and its faith in the invisible hand's ability to transform private vice into public virtue.

Interestingly, the famous invisible hand passage in Adam Smith also occurs in the context of a defense of free trade. Smith takes it for granted that it is in the public interest for national capital to be employed at home, and then goes on to show that, by and large, because of the same community attachments to the home market mentioned by Ricardo, the capitalist will find it in his own personal interest to invest at home. As Smith in his **Wealth of Nations** (p. 423) put it:

"By preferring the support of domestic to that of foreign industry, he intends only his own security; and by directing that industry in such a manner as its produce may be of the greatest value, he intends only his own gain, and he is in this, as in many other cases, led by an invisible hand to promote an end which was no part of his intention."

Smith presupposes that the capitalist is first and foremost a member of the community. Smith's capitalist is so thoroughly English that his very personal identity is defined by internal relations of community with "the country of his birth and connexions". It is not the competitive external relations of the cash nexus that generate the invisible hand in this case, but rather the internal relations of community that constitute this capitalist's very identity. Of course he acts in his self interest, but when the self is constituted by internal relations of community it is not surprising that private interest should promote community welfare. The invisible hand, in its classic statement at least, presupposes the force of community operating within the personal identity of the individual capitalist. By his very self-identity the capitalist feels a "natural disinclination" to invest abroad, and therefore invests at home for his own satisfaction and security, and incidentally promotes the general welfare even though that was not his direct intention.

It is clear that Smith and Ricardo were considering a world in which capitalists were fundamentally good Englishmen, Frenchmen, etc., not a world of cosmopolitan money managers and transnational corporations which in addition to having limited liability and immortality conferred on them by national governments, have now trancended those very governments and no longer see the national community as their context. They may speak grandly of the "world community" as their residence, but in fact, since no world community yet exists, they have escaped from community into the gap between communities where individualism has free reign.

The modern textbook explanations of free trade and comparative advantage make no mention of the assumption of factor immobility. Perhaps one reason this central assumption is downplayed is that frequently the principle is explained in terms of specialization between individuals. A classic example is a lawyer who is a better typist than her secretary. Although the lawyer has an absolute advantage both in knowledge of the law and in typing, she nevertheless finds it advantageous to specialize in law (her comparative advantage) and employ a secretary to do the typing. Since there is no possibility for labor power, human capital, or any other productive capacity to flow out of the secretary and into the lawyer in response to absolute advantage the assumption of factor immobility is guaranteed, and the principle of comparative advantage governs. But the argument cannot be generalized to nations without the explicit requirement that their productive capacities (factors) not flow across national boundaries.

Consider, for example, the dismissal of the (so called) "low-wage foreign labor fallacy" found in one of the best and most widely-used texts (See Lipsey et al., 1987). "...stop and think what the argument would imply if taken out of the international context and put into a local one, where the same principles govern the gains from trade. Is it really impossible for a rich person to gain from trading with a poor person? Would the local millionaire be better off if she did all her own typing, gardening, and cooking?" (p. 795).

In the next paragraph we are assured that "gains from trade depend on comparative, not absolute advantages", which is all very well if capital is immobile. But there is not a word in the discussion about international mobility of factors. Moreover, the shift of context from nations to individuals, far from illuminating the situation, obscures it by abstracting from the very possibility for transfer of productive capacity between the exchanging entities. Economists want free trade, which means that for economic purposes national boundaries are erased. Yet they justify this policy in terms of a doctrine whose central assumption is that national boundaries limiting the flow of capital **not** be erased!

One might object that although the principle of comparative advantage developed historically on the assumption of internationally immobile capital, that does not prove that the assumption is logically necessary, and in fact all one needs to demonstrate the existence of comparative advantage is different relative costs. This is true as far as it goes, but the point is that once capital is mobile absolute advantage takes control, and comparative advantage is superceded as a guiding principle. Capital follows absolute advantage when mobile internationally. Only when capital is immobile internationally will it be reallocated within the nation according to the principle of comparative advantage.

A high wage level is not the only good thing that depends on community. Once community is devalued in the name of free trade and global integration there will be a generalized competing away of all community standards that raise costs of production. Social security, medicare, and unemployment benefits all raise the cost of production just like high wages, and they too will not survive a general standards-lowering competition. Likewise, the environmental protection and conservation standards of the community also raise costs of production and will be competed down to the level that rules in overpopulated third-world countries. Free trade, as a way of erasing the effect of national boundaries, is simultaneously an invitation to the tragedy of the commons. Few people would advocate free migration because they can intuitively see the tragic consequences. Free trade and free capital mobility have exactly the same consequences for wages and community standards, but are widely advocated in the false belief that comparative advantage guarantees mutual benefit.

These arguments should not be taken as advocacy of autarky. The idea is not to oppose international trade, but to suggest that balanced trade between national communities rather than free trade between individuals across national boundaries offers a better context for reflecting and protecting the dimension of community in our economic relations. Individual employees of Exxon cannot trade freely with individual employees of General Motors seeking only their own interest. They must demonstrate that the deals they make are in the mutual interest of the two corporations.

John Maynard Keynes (1933) stated the issue judiciously:

"I sympathize, therefore, with those who would minimize, rather than with those who would maximize, economic entanglement between nations. Ideas, knowledge, art, hospitality, travel - these are the things which should of their nature be international. But let goods be homespun whenever it is reasonably and conveniently possible; and, above all, let finance be primarily national."

Summary and Conclusions

The major conceptual issue we must resolve in thinking about economic development and the environment as the next century begins is to integrate the one-way throughput as a central concept, even more basic than the circular flow, and to distinguish clearly the problem of its optimal scale from its optimal allocation. Our attention will then naturally become focused on how collectively to limit scale to an optimal, or at least sustainable, level. But the community cohesion necessary for such collective action is being eroded by free trade and free capital mobility. Contrary to common opinion the principle of comparative advantages does not solve the problem. New thinking is required.

In conformity with the organization of this conference the focus of this paper has been on concepts, not policies. The specific policy implications remain to be worked out, although the general directions are clear from the above: operationalize the concept of throughput, perhaps in the form of a "biophysical budget" or carrying capacity estimate as a supplement to financial budgets; develop criteria, however rough, of optimal (or sustainable) scale; devise better practical measures of Hicksian income as part of the system of national accounts; protect community from the centrifugal individualism of free trade and free capital mobility, without unduly protecting inefficiency. Translating these general directions into specific country policies is a challenging task. But before making that investment further debate on the general directions is in order, because there is as yet no broad consensus in support of the views just presented. There is, in fact, a substantial consensus to the contrary among economists.

References

Boulding, K. (1966), The Economics of the Coming Spaceship Earth, in: Jarrett, H. (ed.), Environmental Quality in a Growing Economy, Baltimore, Johns Hopkins University Press.

Cobb, J. and Daly, H. (1989), in: For the Common Good, Boston: Beacon Press).

Culbertson, J.M. (1984), International Trade and the Future of the West, Madison, Wisoncisn: 21st Century Press.

Daly, H.E. (1985), The Circular Flow of Exchange Value and the Linear Throughput of Natter-Energy: A Case of Misplaced Concreteness, Review of Social Economy, December.

Georgescu-Roegen, N. (1971), The Entropy Law and the Economic Process, Harvard University Press, Cambridge.

Heilbroner, R. and Thurow, L. (1981), The Economic Problem, New York.

Hicks, J.R. (1948), Value and Capital, p. 172.

Keynes, J.M. (1933), National Self-Sufficiency, The Yale Review, Summer.

Kneese, A.V., Ayres, R.V. and d'Arge, R.C. (1970), Economics and the Environment: A Materials Balance Approach, Washington D.C.

Lauderdale, E. (1891), An Inquiry into the Nature and Origin of Public Wealth and into the Means and Causes of its Increase, second edition, Edinburgh: Archibald Constant and Co., p. 44.

Lipsey, R., Steiner, P. and Purvis, D. (1987), Economics, New York, eighth edition.

Chapter 6

EXTERNALITIES: ECONOMICS & THERMODYNAMICS

R.U. Ayres
A.V. Kneese

Preface

The term "thermodynamics" in the title, and throughout the paper, is used somewhat loosely. Our purpose in this paper is to explore connections between physical constraints (e.g., the "laws" of thermodynamics) and economic phenomena. When we say that a certain physical law implies, or does not imply, some economic relationship or consequence, we mean it in the special sense that when the physical constraint is superimposed on the various relevant economic assumptions or axioms, some conclusion follows. As any thermodynamic purist would point out, the laws of thermodynamics, taken by themselves, have no particular economic implications. We may also offend some purists by not offering formal definitions of numerous terms, such as "the economic system". This is because the exact definition would not in any case affect the argument.

Introduction

A major theme of this paper is the pervasiveness of "common property" problems and market resource misallocations associated with them. This is a consequence of the fact that the real economic system is an open system that must extract, process and discard great amounts of physical materials. All three processes involve significant externalities - market failures-particularly the final disposal of waste residuals. Economic theory has long recognized in a limited way the fact that real markets tend to be imperfect, notwithstanding the great emphasis placed on theorems derivable from the assumptions required for perfect markets. In other words, real markets tend, in practice, to violate some or all of the conditions needed to achieve a Pareto-optimum. A question of some interest to us is whether such an optimum still exists in the absence of complete and perfect "futures" markets for all resources, including environmental resources, and, if so, how it progressively deviates over time from the perfect market case. We argue that, if the dynamic optimum does exist, the current price structure and allocation of resources is quite far from it.

The standard Walrasian (neo-classical) model of an economic system consists of a set of production activities with cost functions, a set of independent resources, commodities and services, and a final demand function. It is assumed for convenience that all resources, commodities and services can be produced entirely from linear combinations of others in the system (i.e., the system is closed), with an appropriate expenditure of labor. The wages of labor constitute the income available to satisfy demand. In the original version, no joint products or co-products are permitted and (by assumption) ther are no "free goods" or wastes. In the 1870s Walras conjectured that such an economy would have an equilibrium set of prices and outputs, toward which the economy would always tend.

The existence of a static equilibrium was finally proved by Arrow and Debreu in 1954 [Arrow & Debreu 54]. This achievement has steered a generation of economists into the analysis of highly abstract mathematical models. Indeed, the rather tight restrictions of the original proof have been significantly loosened. The general equilibrium has even been extended to the dynamic case, with exhaustible resources (e.g. [Solow 74]), subject as before to the assumption of perfect future markets for resources. A great deal of theoretical superstructure has been added on to this basic model in recent years, e.g. [Dasgupta & Heal 79]. To borrow a quote attributed to Milton Friedman, "we curtsey to Marshall but we walk with Walras" [Hoos 83 p. 33].

But, however elegant the Walrasian model may be, it does not represent the real world at all well. It has been properly criticized as "a system devoid of human beings" [Seligman 62, p. 385] and a spuriously deterministic one, at that. Milton Friedman once described it as "a form of analysis without much substance" [ibid], and this criticism remains valid. Moreover, a theoretical objection can be raised: The Walrasian model is essentially closed. It posits abstract resources being converted endlessly into abstract goods and services "consumed" by abstract labor and finally back into resources. This is a kind of perpetual motion machine.

Of course, the word "consumption", so much used in economics, is a slippery and potentially misleading one. Materials and material products (unlike pure services) are not really consumed. The only thing consumed is their "utility". When material objects are worn out or obsolete and useless they are discarded. Fuels and a host of other products that are "used up", are really only dispersed or chemically transformed (e.g., into combustion products). All of the matter extracted and processed each year by the economic system is discarded sooner or later, (but in a thermodynamically degraded form). Only a small percentage of the total mass of materials processed is embodied in long-lived products like books, machines, or structures that last for more than a decade.

As we noted at the outset, the real economy is an open system which utilizes large quantities of material and energy resources provided by nature. These can be extracted, but not "produced" (as intermediate and final goods and services are), by linear combinations of labor, capital goods and other intermediate products. As Georgescu-Roegen has correctly pointed out, the real economy is a dissipative system, not a self-perpetuating one [Georgescu-Roegen 71].

The Walrasian paradigm is fatally flawed not so much by its closure, however, as by its assumption of perfect information availability to all transactors. In fact, a major theme of this paper is the necessity of tehcnological change in open systems, as the highest quality resources are exhausted and replaced by others requiring different technologies. (In an economy with fixed technologies, the mass of raw materials processed and the energy required for processing, per unit mass of final products, would necessarily increase over time even if population and per capita income did not).

However, the direction of technological change cannot be completely foreseen, though some trends are predictable. Even if resource stocks were completely known, there exists a fundamental and irreducible uncertainty, not only with respect to the nature of the next major innovation (say, in energy), but with respect to the secondary and indirect effects of it. Clearly, an industrial system based on solar cells or nuclear power would require very different materials and processes than an industrial system based on petroleum or coalburning steam engines. This uncertainty negates one of the necessary conditions for the existence of a dynamic general equilibrium. "Futures" markets involving technology are inherently imperfect and imperfectible. This means that the kind of economic growth that actually occurs in the real world tends to be discontinuous and otherwise inconsistent with the kind of growth permitted within conventional general equilibrium models of the economy.

The use of energy and matter by the economic system is, in some respects, similar to the use of energy and matter by an ecosystem. As with communities of living organisms, the fraction of each year's mass-energy intake that is actually embodied in the growth increment as new biomass is typically very small compared to the fraction that is used for metabolic activities. The term "industrial metabolism" is a convenient rubric for the economic analog of biological metabolism. We shall return to it later.

Economic Implications of Matter Conservation

The first law of thermodynamics states that the total amount of energy is conserved in all processes, hence in all transformations of materials. Since Einstein postulated the equivalence of matter and energy ($E=mc^2$) it follows that what is conserved is "mass-energy". Matter can, in principle, be created from energy or destroyed to produce energy, but as a practical matter, matter-energy conversion is extremely difficult to achieve under ordinary conditions. Nuclear bombs and nuclear power plants do make use of this equivalence, to produce very large amounts of energy from comparatively tiny amounts of matter. The reverse process - matter creation - is so rare that it is almost unobservable. (It was the dominant process in the universe, however, during the first picosecond after the "Big Bang"). But the essential point is that, with the exceptions noted, mass and energy are conserved independently in most processes on the earth's surface.

As noted above, the real economic system is dependent on the extraction of large quantities of matter from the environment, processing and conversion into various forms, culminating in final products for "consumption". Table 1 (below) shows the mass of "active" raw materials extracted for processing within the boundaries of U.S. for the years 1960, 1965, 1970 and 1975. Vegetable material harvested directly by animals has been omitted for lack of data, along with some obviously minor agricultural and horticultural products. Figures for metal ores exclude mine tailings and gangue removed to uncover ore bodies. Inert construction materials such as stone, sand and gravel have also been omitted. Inert materials account for enormous tonnages, but undergo no chemical or physical change except the extent they are incorporated in concrete or paved surfaces. We also exclude soil and subsoil shifted during construction projects or lost by erosion.

Evidently, more than 10 tons of active mass per person is extracted from its own territory by the U.S. economy, each year. The amount processed is somewhat larger, since the U.S. is now a net importer of many minerals and petroleum. Of course, quite a lot of the grain harvested in the U.S. is exported, which balances the international accounts to some degree. Of the active mass processed each year, roughly 75% is mineral and "non-renewable" while 25% is, in principle, from renewable sources. This point is important when we later come to address the issue of resource prices.

Table 1: Mass of Active Materials Extracted Commercially, U.S. (1960-75) (millions of tons)

Material	1960	1965	1970	1975
Food and feed crops (ex.hay)	267	295	314	403
Meat, fish & dairy products	82	85	85	84
Cotton, wool, hides, tobacco	5	5	4	4
Timber (15% moisture basis)	256	267	271	249
Fuels (coal, lignite, oil, gas)	990	1176	1458	1392
Ores (Fe, Al, Cu, Pb, Zn)	400	435	528	460
Non-metallics	200(e)	240(e)	266	255
Total	2200(e)	2500(e)	2926	2847

It is difficult to estimate the fraction of the total mass of processed active materials that is annually embodied in long-lived products and capital goods (durables). None of the food or fuel is physically embodied in durable goods. Most timber is burned as fuel or made into pulp and paper products. At least 80% of the mass of "ores" is unwanted impurities (more than 99% in the case of copper). Of the final products made from metals, a large fraction is converted into "consumption goods", such as bottles, cans, chemical products, and "throw-away" products such as batteries, light-bulbs, and so on. Only in the case of non-metallic minerals (ignoring inert materials, as before) is as much as 50% of the mass be embodied in durable goods (mainly portland cement used for concrete and clays used for bricks and ceramics). The annual accumulation of active materials embodied in durables, after some allowance for discard and demolition, is probably not above 150 million tons, or 6% of the total. The other 94% is converted into waste residuals as fast as it is extracted. In addition to raw materials, large amounts of energy, in a high quality ("available") form are also needed to drive the system. This energy is also degraded and finally discarded as low temperature heat. Entropy increases.

Figure 1 is a schematic representation of the economy as a materials-process-product transformation system. What it illustrates is that materials processing can be regarded as a series of changes in the energy and entropy states of "useful" materials. But the materials themselves (being conserved) do not disappear after they are "used up" in the economic sense. They become waste residuals that can cause harm and must be disposed of. In fact, it is not difficult to show that the tonnages of waste residuals are actually greater than the tonnages of crops, timber, fuels and minerals recorded by

Figure 1.

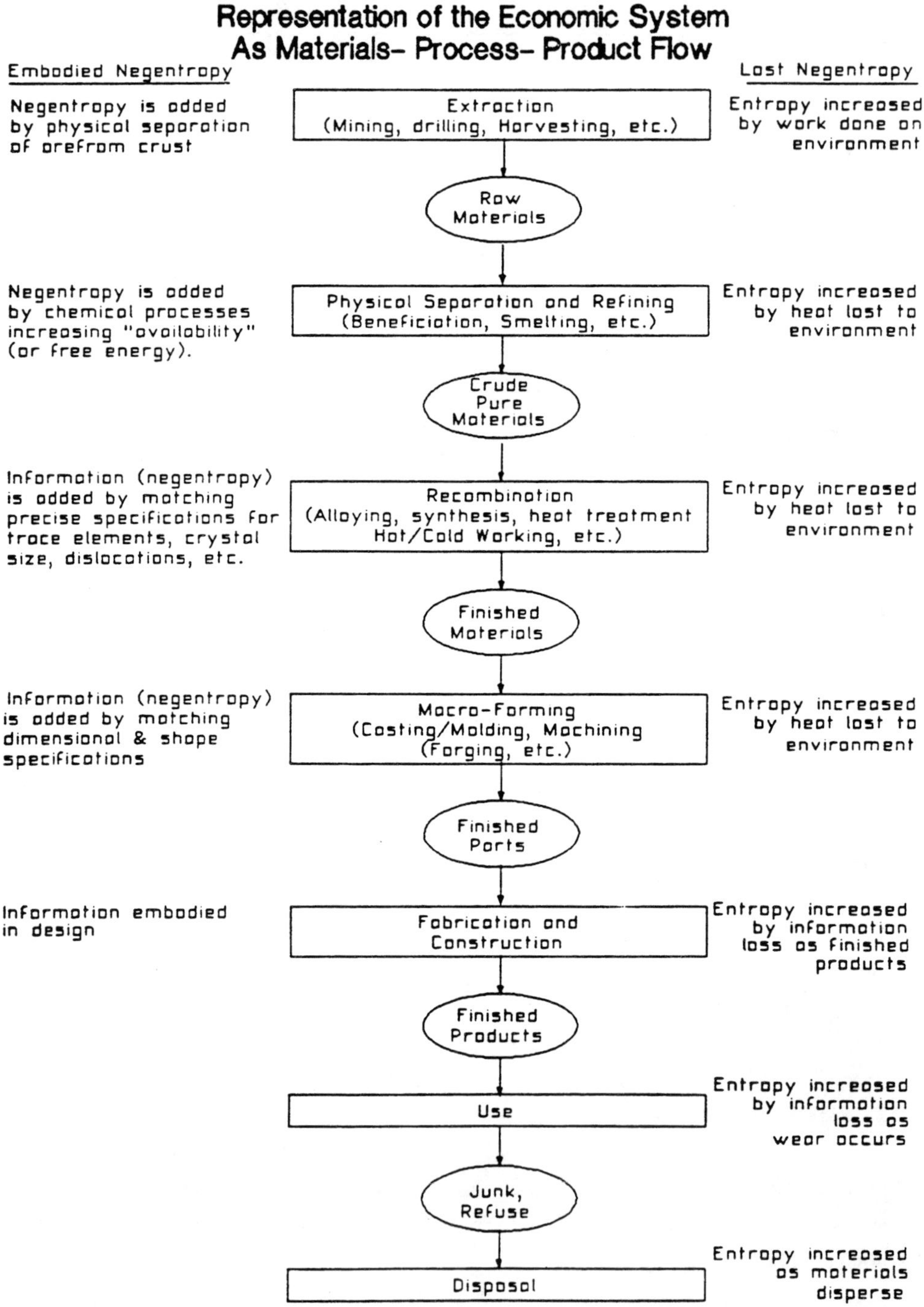

Representation of the Economic System
As Materials- Process- Product Flow
Embodied Negentropy
Lost Negentropy
Negentropy is odded by physicol seporotion of orefrom crust
Extroction (Mining, drilling, Horvesting, etc.)
Entropy increosed by work done on environment
Row Moteriols
Negentropy is odded by chemicol processes increosing "ovoilobility" (or free energy).
Physicol Seporotion ond Refining (Beneficiotion, Smelting, etc.)
Entropy increosed by heot lost to environment
Crude Pure Moteriols
Informotion (negentropy) is odded by motching precise specificotions for troce elements, crystol size, dislocotions, etc.
Recombinotion (Alloying, synthesis, heot treotment Hot/Cold Working, etc.)
Entropy increosed by heot lost to environment
Finished Moteriols
Informotion (negentropy) is odded by motching dimensionol & shope specificotions
Mocro-Forming (Costing/Molding, Mochining (Forging, etc.)
Entropy increosed by heot lost to environment
Finished Ports
Informotion embodied in design
Fobricotion ond Construction
Entropy increosed by informotion loss os finished products
Finished Products
Use
Entropy increosed by informotion loss os weor occurs
Junk, Refuse
Disposol
Entropy increosed os moteriols disperse

economic statistics. While usually unpriced and unmeasured, both air and water are major inputs to industrial processes and they contribute mass to the residuals - especially combustion products. Residuals tend to disappear from the "market" domain, but not from the external world in which the economic system is embedded (Figure 2). Many services provided by the environment are inherently from "common property" - including the air, the oceans, the genetic pool of the biosphere, and the sun itself. Distortions in the market (i.e., prices) are unavoidably associated with the use of common property resources.

One economic implication of the conservation of mass (part of the first law of thermodynamics), is that in an economic system depending heavily on the extraction of non-renewable resources and the use of common-property resources such as air or water, externalities cannot be rare or unusual. On the contrary, as we noted two decades ago [Ayres & Kneese 69], they are pervasive. Furthermore, increased population and higher levels of industrial output based on a fixed technology obviously would require the extraction of increasing amounts of raw materials. This, in turn, would lead to the increasing relative importance of external effects requiring regulatory solutions.

Among the latter, the most promising in the long run are to prohibit dumping of wastes except in carefully monitored locations, to assess the true costs of disposal on users of materials or material products, and to enforce a higher level of materials recycling activity than an unfettered market economy would adopt. The latter makes sense, because recycling simultaneously reduces both the need for extraction - which involves a considerable degree of environmental disturbance - and the need for disposal of waste residuals into environmental reservoirs that are becoming increasingly scarce. Clearly, environmental resources have been significantly underpriced in the past. For this reason, they have been over-used.

A corollary we did not see clearly enough twenty years ago is that the underpricing of environmental resources corresponds to an underpricing of those exhaustible mineral resources whose subsequent disposal as waste residuals causes harm to the environment. (Here the distinction between renewable and non-renewable resources is critical: while renewable resources can obviously create pollution problems, such as sewage, they are almost invariably localized in nature and can be abated at moderate cost. This is emphatically not the case for combustion products of fossil fuels or dispersion of toxic heavy metals, etc.) We argue, in fact, that exhaustible

Figure 2.

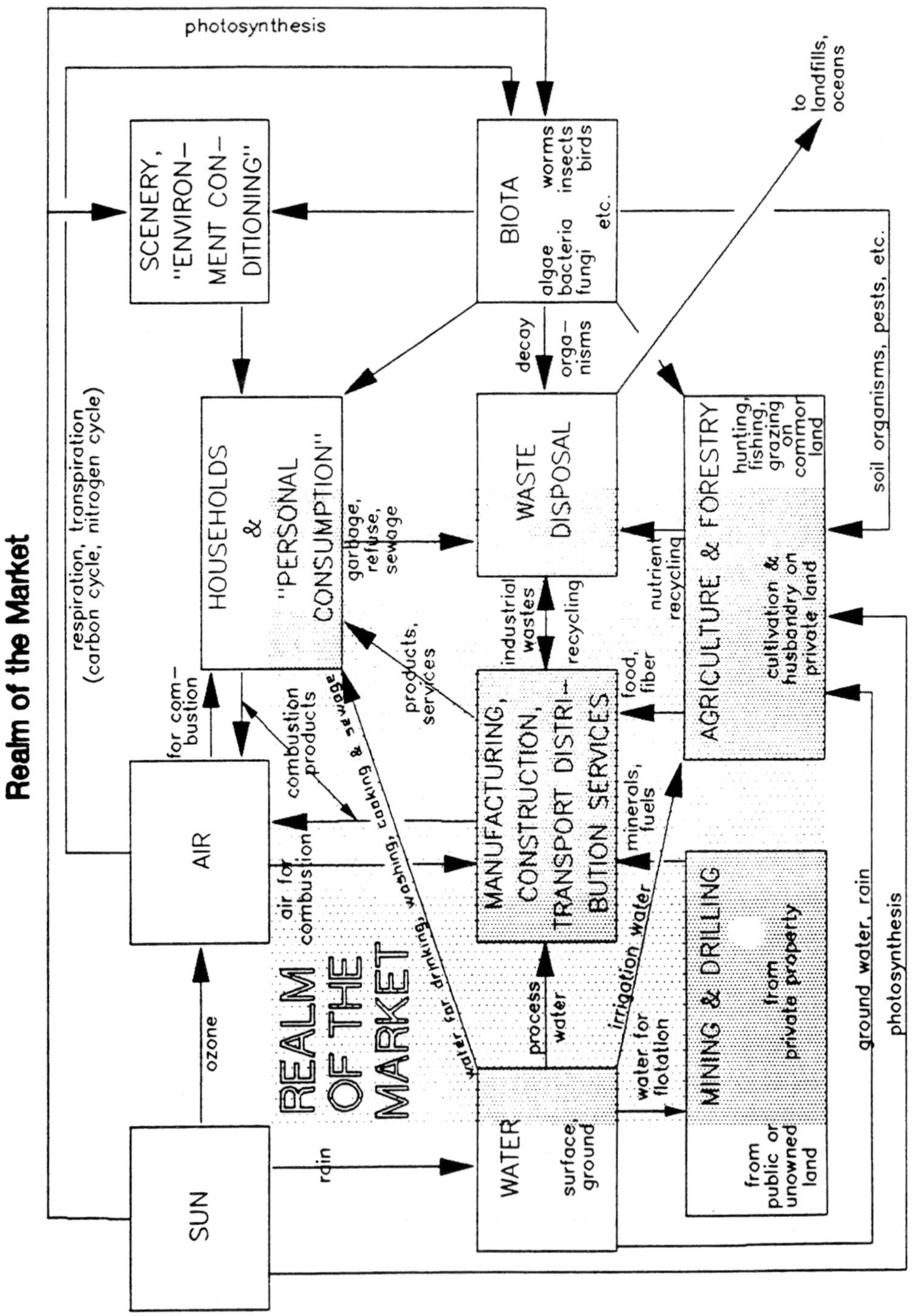
Realm of the Market
photosynthesis
respiration, transpiration
(carbon cycle, nitrogen cycle)
SCENERY,
"ENVIRON-
MENT CON-
DITIONING"
BIOTA
algae
bacteria
fungi
worms
insects
birds
etc.
to
landfills,
oceans
soil organisms, pests, etc.
decay
orga-
nisms
HOUSEHOLDS
&
"PERSONAL
CONSUMPTION"
garbage,
refuse,
sewage
WASTE
DISPOSAL
AGRICULTURE & FORESTRY
hunting,
fishing,
grazing
on
common
land
cultivation &
husbandry on
private land
nutrient
recycling
industrial
wastes
recycling
products,
services
food,
fiber
for com-
bustion
combustion
products
water for drinking, washing, cooking & sewage
MANUFACTURING,
CONSTRUCTION,
TRANSPORT DISTRI-
BUTION SERVICES
minerals,
fuels
AIR
air for
combustion
REALM
OF THE
MARKET
process
water
irrigation water
water for
flotation
MINING & DRILLING
from
private property
from
public or
unowned
land
ground water, rain
photosynthesis
ozone
SUN
rain
WATER
surface,
ground

resources have also been underpriced by a large factor, due to the lack of any link between the market price paid and the subsequent cost of disposal or of uncompensated environmental damage. The true dynamic equilibrium price of any exhaustible resource would have to correspond to a rate of use such that the discounted marginal environmental degradation (cost) resulting from the ultimate disposal of the material is balanced by a discounted stream of marginal benefits. The resource prices that would correspond to a true Walrasian dynamic equilibrium cannot be estimated quantitatively by any "partial equilibrium" approach. However, as an indication of magnitudes, it might be helpful to consider what the cost of energy would be if no incrases in atmospheric carbon-dioxide, or environmental acidity, were permitted. (These, among other constraints, would seem to be implied by the criterion of environmental "sustainability" currently in vogue).

One of the unresolved problems here is the choice of discount rate. While most economists appear to prefer a "market rate of interest", there is a school of opinion (with which we feel considerable sympathy) to the effect that a much lower -possibly zero rate- of discount is the appropriate one. This follows if (on ethical grounds) the interests of future generations are given equal weight with those of the present generation. Taking this position, which has been called "conservationist", no permanent degradation of the environment can be accepted because, without discounting, it results in an infinite cost. From an economic perspective, this is roughly what the notion of a "sustainable environment" implies.

Another important question is the following: Given that the economic system currently depends on processing large amounts of matter derived from outside the system (i.e. from the environment), does "thermodynamics" have anything further to say? In particular, is there any inherent relationship between the amount of matter that must be extracted and the aggregate value of economic services that can be produced? To simplify (perhaps excessively) is there any long-run lower limit to the amount of matter that must be processed to generate a unit of GNP?

The answer is no, both in practice and in theory. The trend toward a service-based economy in the U.S. is well-documented and to the point. Figure 3 shows the rise in employment in information-related activities and the corresponding long-term decline in agriculture, mining and manufacturing [Porat 77]. In general, jobs in the service sectors now contribute more to GNP as jobs in the sectors that process materials.

At the micro-level, too, there ar many examples of dramatic increases in the amount of economic value that can be delivered by a given amount of

Figure 3.

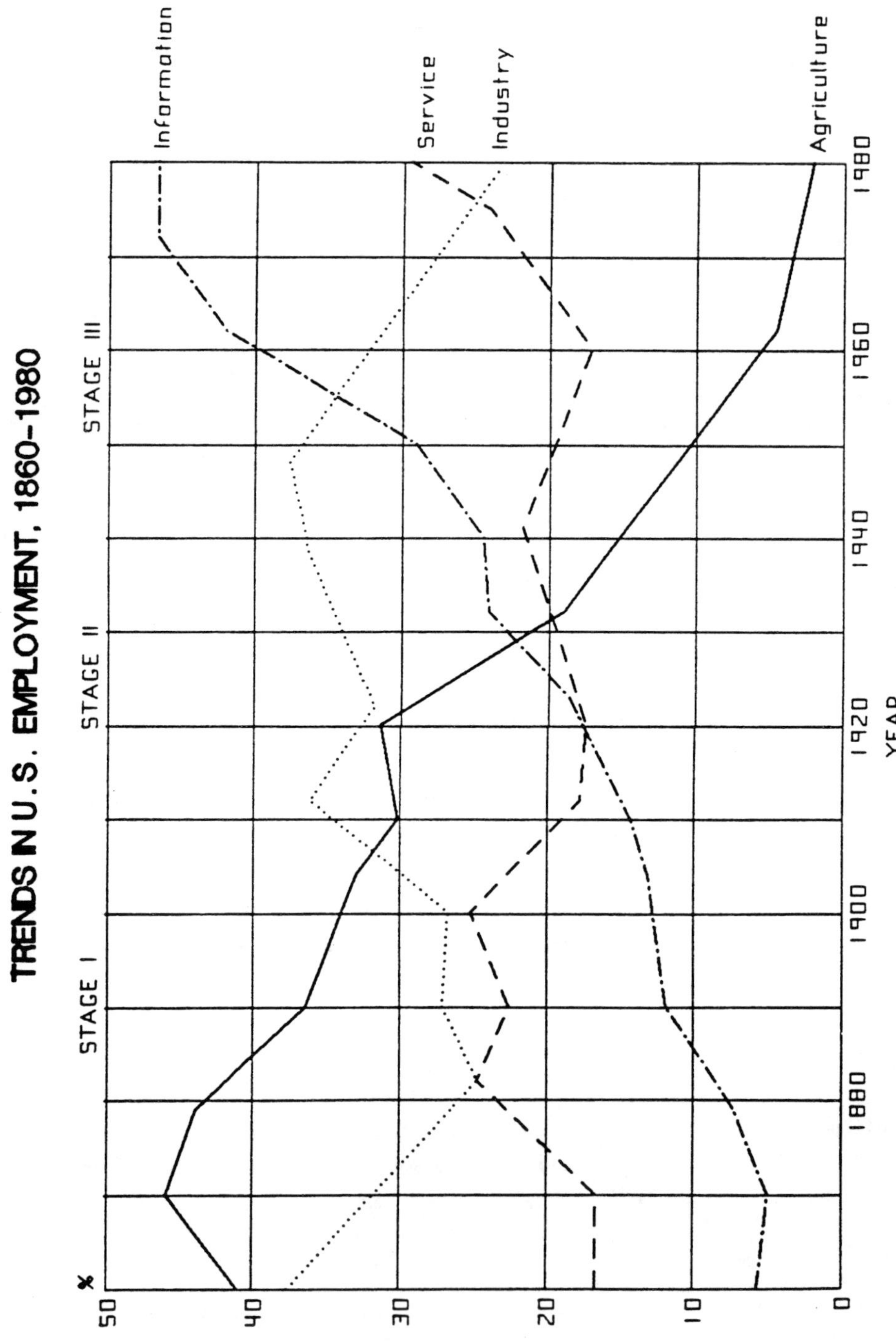

Source: [*Porat 77*]

matter. At the beginning of the industrial revolution, several tons of stone, wood and iron were required to deliver one shaft horsepower via a Newcomen steam engine. Today's aircraft jet engines require less than a pound of materials per (equivalent) shaft horsepower. The shrinkage of computers is even more remarkable. Vannevar Bush's differential analyzer (1935) weighed 100 tons. Though it was an analog computer, its essential functions could easily be performed by a sophisticated pocket calculator today. Yet several models were built nevertheless, to calculate ballistic trajectories, electronic orbits, and the like [Shurkin 85, p. 79]. Eckert and Mauchley's ENIAC (1946), the first purely electronic machine, weighed 30 tons and consumed 174 kilowatts, but it was hundreds of times faster than its predecessors [ibid. p. 139]. One of today's lap-top computers, thousands of times more powerful than ENIAC, weighs only 10 to 20 pounds. Yet the trend to miniaturization apparently has much farther to go, as functions now carried out by thumb-nail sized silicon "chips" may eventually be performed by individual custom-grown organic molecules.

The theoretical reason why physical considerations do not imply a lower limit to the Extracted Matter per unit of GNP (EM/GNP ratio) is that as long as energy is available there is no fixed upper limit - short of 100% - to the fraction of materials that can be recycled. In other words, energy is the only limiting factor. This point will be taken up again later in connection with the "second law", where it is more relevant.

Implications of Conservation of Energy

We have noted above that the "first law" of thermodynamics is better known as the law of conservation of energy. Given that the economic system is driven by a flux of thermodynamically available energy that must be provided from outside the system, and (as will be seen later) cannot be recycled, the corresponding question arises: Is there any inherent relationship between the amount of available (or free) energy that must be supplied and the economic value (GNP) that can be generated? Again, the answer is no. Again, we can argue the point on both empirical and theoretical grounds.

At the macro-level, the record is somewhat confusing at first glance. What it seems to show is an increase in energy-use per unit of GNP during the industrialization stages of development, followed by a decline as countries enter the "post-industrial" phase, with increasing contributions from the service sectors and, particularly, the information-intensive activities. In the case of the U.S. the period of increasing Energy/GNP ratio ended in the 1920s, and there has been a long-term decline since then, as shown in Figure 4. From 1929 to 1955 the decline was 24%. [Schurr et al 60, p. 177]. However, this decline was interrupted by a period of relatively constant ratio lasting from about 1950 until 1967. (An increase of 7% from 1968 through 1970 was probably due to the Viet Nam War). Real energy prices in the U.S. actually declined fby 10% from 1953 to 1973 [Dunkerley 80 p. 56]. After 1973 energy prices rose sharply and the long-term decline in E/GNP continued. There was a further 21% drop from 1976 through 1984.

The experience of other countries has been qualitatively similar, but peak E/GNP levels have been reached more recently. France and West Germany both peaked in 1957 [ibid], while Sweden, the Netherlands and Japan all reached their peaks in the mid-70s. The OECD as a whole experienced a drop in E/GNP ratio of 19% from 1973 to 1984. For developing countries, the ratio is still growing. Nevertheless, the record strongly suggests that E/GNP ratio's will decline once the industrialization process is well advanced.

Yet studies of the last three decades (for which reasonably good data are available for many countries) have also lent unjustified credence to the notion that the link between energy and GNP might be quite inelastic. A number of studies in the early 1970s suggested that energy supplies could not be curtailed without drastic reductions in standards of living. It was also suggested repeatedly by reputable analysts that the maximum potential for "conservation" was only of the order of 20%. For example, in 1977 the OECD estimated maximum potential conservation possibilities as follows [OECD 77]:

Energy sector	1%
Industry	4%
Transportation	7%
Residental/commercial	4-5%
Total	16-19%

This "maximum" saving was exceeded well within a decade.

The widespread assumption by economists that the E/GNP ratio is inelastic is parallelled by a pervasive misunderstanding of the distinction between energy use, and efficient energy use, to which engineers have also contributed. This misunderstanding amounts to a myth, viz. that roughly 50% of energy inputs are "used" by our economy (presumable efficiently)' while the other 50% is "lost". This summarizes the message conveyed by a very widely ciruclated pamphlet entitled "Understanding the National Energy

Figure 4.

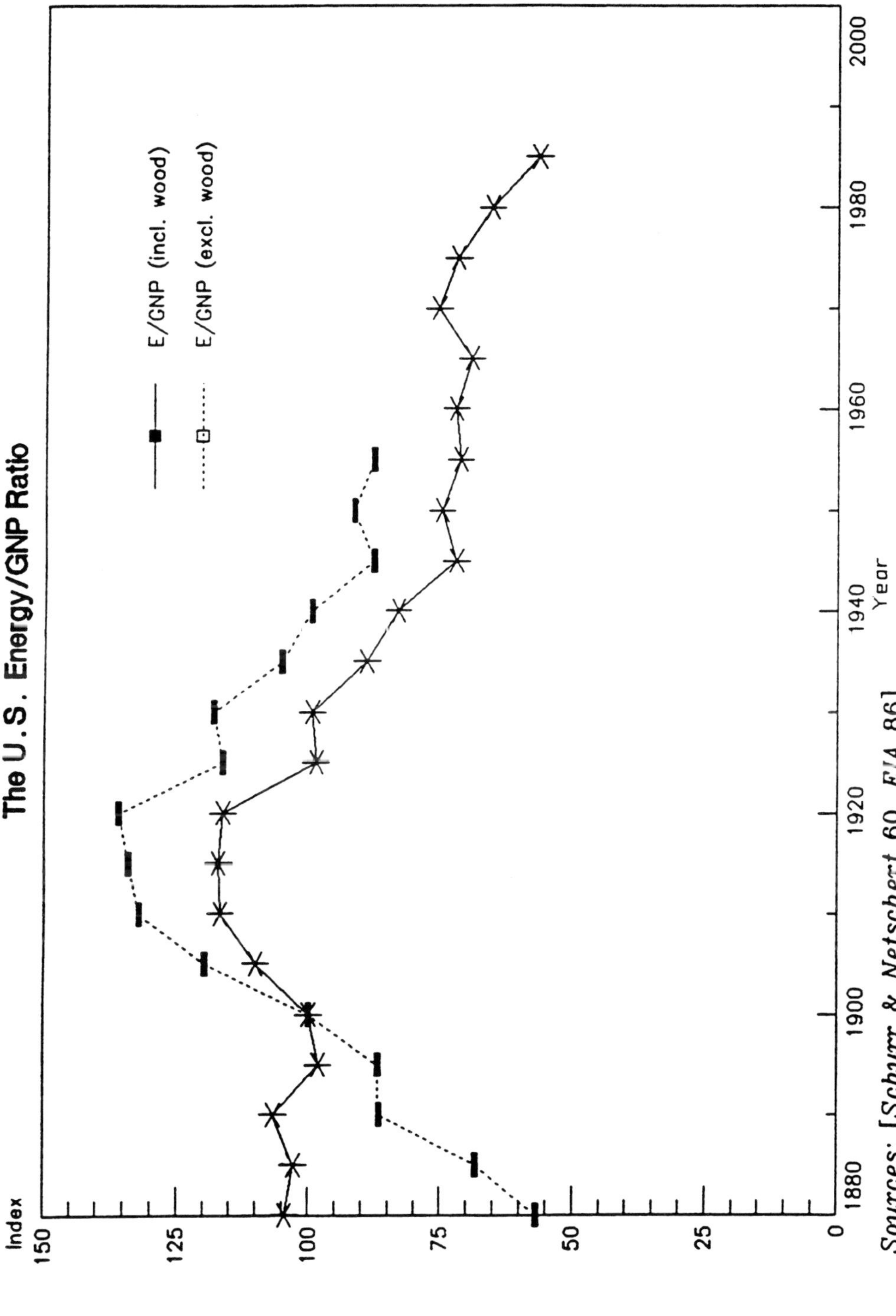

Sources: [*Schurr & Netschert 60, EIA 86*]

Dilemma", prepared with the help of Livermore National Laboratory and the staff of the Joint Committee on Atomic Energy [Bridges 73], and circulated by the JCAE. An updated version, entitled "Energy: A National Issue" was even published by the Center for Strategic and International Studies of Georgetown University [Murray 76].

According to the latter version, aggregate energy "loss" rates for various sectors were as follows:

Electricity Generation & Distribution	65%	(implied efficiency 35%)
Industry	30%	(implied efficiency 70%)
Transportation	75%	(implied efficiency 25%)
Residential/Commercial	30%	(implied efficiency 70%)
Total	51%	

If it were true, as these reports (and many others) suggested, that only 51% of all energy inputs to the economy were wasted, it must follow that the economic "engine" is already using energy with an average overall efficiency of 49%. If this were so, it also follows that very little more energy could be saved by conservation.

The above conclusion is wrong and, not to mince words, absurd. In the first place, it is based on a technically incorrect concept of thermodynamic efficiency. This was pointed out almost immediately in the report of a summer study supported by NSF and sponsored by the American Physical Society [Ross et al 75]. This report was later reprinted as an appendix to some congressional hearings, but had little impact. Yet, based on carefully documented calculations, the authors showed that the true thermodynamic efficiency of burning oil or gas in a furnace to supply warm air at a temperature of about 70° F to a room (the major use of energy in the residential/commercial sector), far from being 70%, is no more than 3%! Allowing further for the fact that many buildings are underinsulated (i.e. they leak heat), and many rooms are overheated, or are heated even when there are no occupants, the real efficiency of the current technology for space heating is much closer to 1% than 70%. Consequently opportunities for conservation are almost unlimited, a fact that is beginning to be cautiously recognized by scientists, at least [e.g. Rosenfeld & Hafemeister 88].

Though we cannot present additional details here, it must be said that the same basic conclusion can be reached for all of the other economic sectors. On an overall basis, even allowing for somewhat higher real efficiencies in the basic industrial sector, the overall efficiency with which the U.S. used energy in 1975 was probably less than 5% and perhaps considerably less [Ayres & Narkus-Kramer 76][Ayres 88]. The above figures do not reflect the possibility of fundamentally new technologies or new kinds of services requiring very little energy to deliver. Opportunities for conservation are correspondingly great. In terms of the E/GNP ratio, it seems likely that a fourfold reduction (or more) could be achieved even without inventing "radical" new technologies or extraordinary costs. Fuller documentation of this argument wil have to be presented in another paper, however [Ayres, 1988].

Implications of the Second Law: No Perpetual Motion

The second law of thermodynamics states, in essence, that "entropy" (a certain property of systems, which we do not propose to define rigorously here) increases in any irreversible process. A closed system is one that exchanges neither matter or energy with its surroundings. Indeed, except in exceptional and practically unrealizable situations, all real processes are irreversible, hence total entropy increases in virtually every real process. When entropy increases the energy in the system becomes less available to do "useful work". In engineering practice, entropic calculations are primarily useful for purposes of analyzing the efficiency of alternative energy-conversion systems and chemical reaction paths. In this domain, there are important economic tradeoffs between reaction rates, yields and efficiency, for example [Tribus & El-Sayed 81], [Kenney 84].

Most non-physicists are more familiar with simplified versions and (apparent) implications of this law. (One example of the latter is the so-called "heat-death" (Wärmetod) of the universe, which was popularized at the end of the 19th century in the context of debates about whether biological evolution is "contrary" to the laws of thermodynamics). A simple implication of the law is that all energy conversion systems and matter conversion systems require more available (or "free") energy inputs than they yield as outputs. In other words, all biological and industrial processes are less than 100% efficient. In still simpler terms, it is impossible to construct a perpetual motion machine - even an abstract one like the Walrasian economic system.

It follows from the second law, incidentally, that no material recycling process can be 100% efficient. A seemingly natural implication of this true statement is that stocks of matter must therefore be dissipated gradually, just as available energy is. Indeed, Georgescu-Roegen has reached this conclusion. Unfortunately he has elevated it into a "fourth law" of thermodynamics [Georgescu-Roegen 77,79]. If Roegen's "4th law" were valid, it would follow that even a continuous flux of solar energy would not be enough to ensure the possibility of a "sustainable" (stationary-state) economy. The earth would eventually "run down" like a mechanical clock, as its stocks of high-quality material resources (such as metal ores) were used up. Given the rate at which "grade deflation" for ores like copper seems to be occurring, the exhaustion-point would seem to be nearer a few hundreds of years from now than a few billions of years. Indeed, the sensational Club of Rome Report "Limits to Growth" assumed that reserves would be exhausted even sooner [Meadows et al 72]. The simplistic world-model presented in that book, and the whole notion of world-modelling, has been largely discredited since then.

Faced with a neo-Malthusian ("fixed resource base") argument, economists have always stressed the potential for technological substitution [Goeller & Weinberg 76], [Tilton 83]. For example, if copper becomes scarce, we switch to aluminum conductors; if tungsten becomes scarce, we can use molybdenum for tool steel; if cobalt and chromium are unavailable for "superalloys", we will develop ceramic turbine blades, and so on. Many such substitutions have already taken place and more will occur in the future. But this does not really contradict Georgescu-Roegen's "4th law" of matter dissipation. It merely delays the day when some "critical" material becomes unavailable and cannot be replaced by another. (For example, there is no technological substitute for phosphorus as a plant nutrient). In the long run, substitution is no substitute for at least some recycling.

Fortunately for the prospects of mankind beyond the next few centuries, there is a subtle but critical flaw in Georgescu's gloomy argument about the impossibility of recycling. It is perfectly true that no industrial process - including recycling - will ever be 100% efficient. But the earth is not a closed system as long as the sun shines, so energy availability is not a problem. Nor is it necessary to achieve thermodynamic perfection to separate useful materials from the terrestrial environment in sufficient quantities to operate an industrial economy. All that is needed is (1) enough solar energy and (2) a gravitational field strong enough to prevent matter from escaping into interstellar space. Given that the earth is a closed system with respect to matter (but not with respect to energy), and the fact that matter is conserved in a closed system, the most unfavorable limiting case is that one might have to extract materials from a "soup" with the average composition of the earth's crust.

In fact, a moment's reflection is enough to see that things can never get even that bad, by a considerable margin. Interatomic forces differentiate materials by chemical properties, such as reactivity, solubility, freezing/melting point and vaporization temperature. At the temperature of the Earth's crust some elements are gaseous, some compounds are liquid and some are solid. Some solids dissolve in water, others don't. The ocean will always have a much higher concentration of soluble salts (e.g. sodium, potassium and magnesium chlorides, bromides, etc.) than the crustal average while the atmosphere will always have a much higher concentration of other elements (such as oxygen, nitrogen and the rare gases). Natural photosynthetic processes also reconcentrate some elements, notably carbon, while hydrogeochemical processes reconcentrate others, including phosphorus, sulfur, calcium, manganese, copper and cobalt. In short, there is no danger of actually running out of materials, even the scarcest ones. A stationary-state economy will, however, have to utilize very low grade sources, in some cases, to replace whatever is "lost" through imperfect recycling. Some of the technical implications of this will be discussed later.

To summarize the last few paragraphs, the second law of thermodynamics does not imply (as argued by Georgescu-Roegen) that a sustainable, stationary-state economy is impossible on the earth, - at least as long as the sun continues to provide a flux of available energy (essergy). To be sure, the long-run prospects for economic growth (and human survival) are thermodynamically constrained, but no absolute limits are apparent from here. Thermodynamic considerations can also be integrated formally with traditional economic concepts to yield new insights into optimality conditions for resource and environmental management. In particular, the problem of optimizing resource extraction policy given a declining grade of resources can be treated more realistically (e.g. Faber et al 87).

Entropy, Self-Organization and Walras

The economy is like a living organism to the extent that it exhibits the attributes of self-organization and growth. For any real physical system to possess such attributes, two fundamental conditions are necessary. First, the system must be open and must receive an influx of available energy (e.g. sunlight) from outside. Second, the system must be non-linear and far from thermodynamic equilibrium. (Incidentally, it must be noted that self-organization cannot be attributed to the second law, although it can be described in thermodynamic language. It is a consequence, if anything, of the non-linearity of nature. It appears quite likely that non-linearity, in turn, is the root-source of irreversibility).

The words "far from equilibrium" require some explanation. In brief, thermodynamic equilibrium is that state of maximum uniformity and "randomness" (or entropy) toward which all spontaneous processes seem to tend, at least when they are fairly close. Near equilibrium, the equations that describe the relationships of state variables can always be linearized. However, far from equilibrium, the state equations of real systems are always non-linear [Nicolis & Prigogine 77]. One can define "far from equilibrium", then, as being far enough for the linear approximation to break down. Or, to put it another way, "far from equilibrium" means in the non-linear region of the equation of state. (In this region, the notion of entropy is not even necessarily well-defined, although it can sometimes be used).

An important attribute of non-linear equations must now be noted: they often have multiple solutions. In fact, as one moves away from the linear region, where there is a unique solution, the solution may "bifurcate" and new possible solutions can exist that do not evolve continuously into the limiting case. The physical configurations of matter that correspond to such alternate solutions of the equation of state typically exhibit a kind of order and persistence which has been called "self-organization" by Prigogine and others. Examples of self-organization can be found in most fields of science. The vortices that form in turbulent fluids, from sun-spots to hurricanes, are familiar examples.

Every living organism is far from thermodynamic equilibrium in the relevant sense, and is maintained by a flow of available energy from the outside. The same is true of the biosphere. Life on earth is energized by the sun. Entropy presumably continues to increase in the universe as a whole, but if one measures the entropy changes within a living organism, the reverse is true. Entropy within the (open) system boundaries actually decreases. This is reflected in the accumulation of energy-rich material (e.g. carbohydrates, proteins and fats) and, more subtly, in the organization and structure associated with life processes. This local decrease does not violate the second law of thermodynamics at all. It is just that the system in question is not a closed one. That it is also "far from equilibrium" is clear from the rapidity with which living organisms break down after death.

The human socio-economic system is evidently a self-organizing system. (The economic part of the system in question can be defined as the sum total of all human activities concerned with producing goods and services, irrespective of whether there exists a free market for them). It is clearly not a closed system, since materials are constantly cycling through it. It requires a flux of available (free) energy. Before the industrial revolution, most work expended in economic activity was derived from human muscles or those of animals. Ultimately, the source was the sun, whose energy was captured by green plants and thence cycled through the food-chain. Today, more than 90% of the energy utilized by industry is derived from fossilized hydrocarbons representing accumulations from millions of years of photosynthesis. Clearly, we are expending this resource at a fairly rapid rate, and if it continues the stockpiles we are using will be exhausted in a century or so.

Similarly, as noted already, we are using up and irreversibly dissipating the best quality ores of iron, copper, lead, nickel, chrome and a number of other metals. To adapt to this decrease in resource quality and availability, a number of major changes in our self-organizing economic system must occur. The "industrial metabolism" of the future will differ from that of today, as the present differs markedly from the past. More energy will be needed for extraciton and refining, deeper mines and wells will be dug, new separation processes will be introduced, more recycling will be done, substitutions will take place. Only the technological details are in doubt.

Incidentally, the present arrangement for disposing of waste materials is equally unsustainable, because it depends largely on dilution rather than recycling, and the dilution factor or absorption capacity permitted by land, water and air resources is, or soon will be, inadequate. As Kenneth Boulding pointed out very cogently some years ago, the days of the "cowboy economy", where human activity could be regarded as an insignificant disturbance to a natural environment of near-infinite capacity, must be replaced by a "space-ship" mentality, in which the earth is regarded as a finite - even small - place [Boulding 66]. This limitation on environmental dilution/ absorption capacity has already reached the critical stage with regard to certain wastes, such as trash, toxic chemicals, chlorocarbons and carbon dioxide. Again, the economic system will continue to be self-organized, but the organization of our industrial metabolism must change.

What is not possible for our economic system, notwithstanding its self-organizing character, is to achieve a sustainable stationary state based on a continuation of the present technology of production and industrial structure. This is not an argument for or against particular political or institutional arrangements (e.g. capitalism vs. socialism). It is a plain statement that the input-output table describing the industrial metabolism of 2080 must look very different from the current one because different resources, requiring different technologies, will necessarily be used.

Incidentally, it is perhaps worth pointing out that, even if we could describe the technology of the (hypothetical) stationary state of the distant future - with its total dependence on solar energy in one form or another, its highly developed recycling processes for materials of all kinds, and its sophisticated methods of extracting metals and chemicals from extremely low grade sources (such as ocean water), - it would be economically irrational to adopt those technologies today. The costs would be excessive in relation to our current productive ability, and probably far beyond our present ability to pay. The optimum policy with respect to any exhaustible resource, as Herfindahl pointed out [Herfindahl 67], is to exploit the highest quality resources first and reinvest the profits to build up the capital and knowledge stock to permit lower grade resources to be used later. In effect, high quality resources in the ground are a form of capital. This capital can, of course, be invested wisely or squandered in excessive consumption. That is our choice.

In short, the most important economic implication of the second law of thermodynamics - as shorthand for the fact that the economic system is open, self-organizing and inherently dissipative, depending on external sources of free energy, raw materials and environmental resources - is the necessity for continued technological change and adaptation. This implies the impossibility of achieving a Walrasian equilibrium based on fixed technologies. At present we are depending on exhaustible stocks of all these resources. In the very long run the exhaustible high quality resources will run out (although the constituent atoms will not) and only the sun's energy will remain available to us. By that time a completely new and different industrial metabolism will necessarily have replaced the current one.

Although this paper is too short to go into details, we cannot resist pointing out again that the prevailing neo-classical economic model is, at bottom, predicated on perfect knowledge of all resource stocks and all technologies. Growth is permitted by the standard model, but technological "surprises" are not. Once the exhaustible resources are gone, the only kind of economic growth that is consistent with that model is growth of all sectors in exact proportion to each other. In biology, this kind of growth is seen in populations of bacteria or coral reefs, but not of complex organisms, where growth is inevitably accompanied by morphological change ("ontogeny recapitulates phylogeny"). The economic analog of a coral reef might be an economy of small, independent, self-contained farms. Needless to say, it does not exist today, if it ever did.

Implications for Industrial Metabolism, Past, Present and Future

As noted above, the total mass of residuals produced each year is likely to exceed the mass of active inputs. This is because nearly half of the inputs are fossil fuesl (hydrocarbons), which combine with atmospheric oxygen and form carbon dioxide and water vapor. The carbon fraction of hydrocarbons ranges from 75% in methane to about 90% in coal. Petroleum is intermediate. The average for all fossil fuels cannot be less than 82%, which translates into around 1200 tons of carbon in 1970. Imported oil would have increased this figure somewhat. In inefficient combustion processes up to 10% of this carbon is converted into carbon monoxide, while carbothermic reduction of iron ore and other metals generates even more, but most carbon monoxide is converted to carbon dioxide at the point of generation. Gross emissions of CO to the atmosphere in the U.S. were about 110 million tons in 1970 (mostly from automobiles and trucks), with a carbon-content of 47 million tons, or about 3.5% of the total. Emission controls reduced this to about 85 million tons net in 1980 [EPA 86].

Carbon monoxide is, of course, quite toxic to humans and has been implicated in health problems among urban populations. It is less well known that carbon monoxide plays an active, and not necessarily benign, role in a number of atmospheric chemical reacitons. Eventually, of course, all carbon monoxide oxidized to carbon dioxide. Thus, 1 billion tons of fuel carbon is converted ino 3.67 billion tons of carbon dioxide emitted to the atmosphere. (Note that the tonnage of carbon-dioxide emitted annually exceeds the total tonnage of active material inputs!) Not all of this remains in the atmosphere, but enough does to raise very serious concerns about the long term climatic impact of continued fossil fuel combustion. The problem is the "greenhouse effect". It appears that the carbon dioxide level of the earth's atmosphere has risen over the last century from about 290 ppm to around 340 ppm at present. While the baseline and the climatic impacts of the rising level of carbon dioxide are both still uncertain, it is already clear that "industrial metabolism" is influencing the natural environment in ways that appear increasingly adverse.

The discovery of chlorofluorocarbon gases in the stratosphere, here they can be ionized by the ultra-violet (U.V.) radiation of the sun, together with the ability of the chlorine ions to scavenge stratospheric ozone by a catalytic reaction, has raised an even more frightening prospect: ozone depletion. Apparent confirmation of this phenomenon has recently appeared in the form of an "ozone hole" in the stratosphere over Antarctica. Chlorofluorocarbons are industrial gases, discovered in 1928 and produced since the 1930s, whose inertness was once thought to be a great virtue. They are used commercially as refrigerants, solvents and for "blowing" plastic foams. In most uses they are not released deliberately, but losses and leakage are inevitable. Precisely because they do not break down in the presence of oxygen, they have now dispersed throughout the atmosphere. The problems revealed so far may only be the beginning. If (perhaps it might be better to say "when") the ozone level in the stratosphere is depleted, the effect will be to let more of the sun's U.V. radiation through to the earth's surface. The ecological impact on vulnerable species is incalculable, at present, but could be extremely severe.

Methane, oxides of nitrogen and sulfur oxides are other residuals that have been seriously implicated in climatic and/or ecological effects. All three are generated by fossil fuel combustion, as well as other industrial processes. Like carbon monoxide, carbon dioxide and the chlorofluorocarbons, they can be considered as metabolic products of industry. Methane is lost in natural gas pipelines, petroleum drilling, coal mining and several kinds of intensive agriculture (especially rice cultivation and sheep farming). Sulfur oxides are generated by combustion of sulfur-containing fuels-especially bituminous coal - and by the smelting of sulfide ores. Most copper, lead, zinc and nickel ores are of this kind. In principle, sulfur can be recovered for use from all these activities, and the trend is rising (Figure 4). But the costs of recovery, especially from coal-burning electric power plants, are far higher than the market value of the products (e.g. dilute sulfuric acid) and, for the present, constitute a waste residual that must be disposed of. This will only change if (or when) the economic value of sulfur rises very sharply above its present low level. This, in turn, depends on the discovery of new uses.

Nitrogen oxides are also co-products of combustion. In effect, at high temperatures, some of the atmospheric nitrogen is literally "burned". Again, in principle recovery for use (e.g. as nitric acid) is possible. But in comparison with the commercially dominant process of nitrogen fixation based on natural gas (the Haber-Bosch process), there is no exonomic justification for such recovery. It is only because dispersion of oxides of nitrogen is implicated in smog and acid rain that recovery technology is even being explored. Only a spectacular rise in the economic value of nitric acid would change this picture. Only new technology could lead to such a change.

Fly-ash is yet another by-product of combustion. At present it is being recovered fairly efficiently from stack-gases of large utility boilers and industrial furnaces, by means of electrostatic precipitators. But the ash itself has become a large-scale nuisance because there exists no use or market for it. The amounts are large: over 50 million tons per annum in the U.S. alone. At least two possible remedies exist, however. Fly-ash is a potential "ore" for several metals, especially iron, aluminum and silicon. These could probably be recovered commercially if, for example, bauxite became unavailable [Ayres 82b]. Alternatively, fly-ash could be used as a substitute, or more likely as a supplement, for portland cement in the manufacture of concrete and concrete-like products. Its major disadvantage in this application is that concrete made with fly-ash does not harden and set as rapidly as the commercial variety. This has obvious economic costs, but so does the disposal of fly-ash into land-fills. A final possibility worth mentioning is that fly-ash could be used as a medium for the disposal of toxic liquid wastes, yielding a hard, impermeable substance suitable for long term storage. This technology is already being developed in France.

Speculations along the above lines are not as "far out" as today's industrialists might believe. Several of the most valuable products of the chemical industry today began as unwanted by-products. Coal tar was one of the first. It became a valuable source of chemicals such as aniline dyes in the 19th century. Coke oven gas became not only a valuable fuel, but also the first commercial source of ammonia. Slag from blast furnaces was once a nuisance. Today it has a number of major uses and old slag heaps are even being "mined" in some locations. Natural gas was once an unwanted by-product of petroleum drilling - but not for long. Chlorine is a less well-known example, being originally a by-product of the manufacture of caustic soda (lye) by electrolysis of brine. Today chlorine is the primary product. Much of the petro-chemical industry is built on by-products of the natural gas industry: ethane, propane and butane. Helium is also taken from natural gas. Hydrochloric acid and sulfuric acid are increasingly being obtained as by-products, and many other examples could be given.

It is probably safe to say that the industrial metabolism of the next century will recycle a number of the major tonnage waste products of today, notably sulfur, fly-ash and lignin wastes from the paper industry. The latter may yet turn out to be a useful growth medium for single-cell organisms providing high-protein supplements for food products.

Combustion and carbo-thermal reduction processes are the major sources of atmospheric pollutants today, but by no means the only important ones. Nor is the atmosphere the only vulnerable part of the environment. From a broader environmental perspective, the production and dispersal of thousands of synthetic chemicals - many new to nature, and some highly toxic, carcinogenic or mutagenic - and the mobilizaiton of large tonnages of toxic heavy metals may be of equally great concern. The complexity of the problem is too great to permit any kind of short summary. However, two points are worthy of emphasis. First, as noted above, most material "pass through" the economic system rather quickly. That is to say, the transformation from raw material to waste residual takes only a few months to a few years in most cases. Long-lived structures are very much the exception, and the more biologically "potent" materials are least likely to be embodied in a long-lived form.

The second point, emphasized strongly some years ago [Ayres 78] but still worth repeating, is that many materials uses are inherently dissipative. That is, the materials are degraded, dispersed and lost in the course of a single normal usage. In addition to food and fuels (and additives such as preservatives), this applies to packaging materials, lubricants, solvents, flocculants, anti-freezes, detergents, soaps, bleaches and cleaning agents, dyes, paints and pigments, most paper, cosmetics, pharmaceuticals, fertilizers, pesticides, herbicides and germicides, and so on.

On reflection, most such uses (food and fuel again excepted) are generally seen to be non-essential in the sense that technologies are theoretically available, or imaginable, that could eliminate the need for them. To take one example only: hydroponic agriculture in enclosed, atmospherically controlled greenhouses, with genetically engineered "anti-pests", would eliminate all losses of fertilizers and pesticides to water-courses by way of surface runoff. It is obvious, however, that the techno-structure of a society employing such technologies would be very different from the present one. The developmental path "form here to there" involves fundamental structural changes not at all compatible with simple expansionary growth along an equilibrium "turnpike".

We conclude with a final point of a more methodological nature. It is that the "materials-balance" principle, a straightforward application of the first law of thermodynamics (widely used in the design of chemical engineering systems, for instance), is a potentially valuable and under-utilized tool for using economic data in environmental analysis. Not infrequently a combination of input data (obtainable from economic statistics), together with technical process data available from engineering analysis, gives a more reliable estimate of waste residual outputs than direct measurements could be expected to do.

One good example of this arises in the aluminum industry. One of the major environmental problems associated with aluminum smelting in the past was the emission of gaseous fluorine from the smelter. The source of the fluorine is the electrolytes (molten cryolite and aluminum fluoride) used as a solvent for alumina in the electrolytic cell. An unavoidable side-reaction in the cell breaks down these electrolytes and releases some of the fluorine at the anode. Exact "recipes" for the production of aluminum are only known by the aluminum companies, but a materials-balance analysis for the year 1973 suggests that for each 100 kg of aluminum produced, 2.1 kg of cryolite and 3 kg of aluminum fluoride were consumed as inputs [Ayres et al 78]. Based on these numbers and known chemistry the aluminum industry would have accounted for 40% of the known production of hydrofluoric acid in that year - consistent with both official (Dept. of Commerce) and unofficial estimates [ibid]. Note that, in the absence of recovery facilities all of this fluorine must have been emitted to the air, since consumption of cyolite and aluminum fluoride is exclusively to replace fluorine losses. It is interesting to observe, also, that the imputed emissions were about twice as high as EPA's published estimates based on direct (but unreliable and difficult to verify) measurements. Assuming the production and use statistics for hydrofluoric acid were correct, one would have to believe that the indirect estimate is more reliable than the direct one.

Another application of the materials-balance methodology is in the reconstruction of historical emissions data. This is a problem of some importance to basic enivironmental science, since the cumulative impact of air or water pollution over long periods can only be evaluated in relation to a baseline of some sort. In this context it becomes important to know more about emissions in the past, when no measurements were made. To be sure, sediments and ice-cores offer some help, but not enough. The picture can be clarified considerably, however, with the help of synthetic models utilizing production and consumption data (which is often imperfect, but better known than emissions) together with engineering analysis of processes. Sometimes process information is not even needed.

For example, fairly good historical estimates of sulfur oxide emissions (SOX) - needed to analyze the long-term impact of acid rain, among other things - can be reconstructed quite easily from historical statistics on coal consumption and copper/lead/zinc smelting [e.g. Gschwandtner et al 83]. This is because the sulfur content of coal and metal ores can be assumed to be the same in the past as it is today, and until recently all of that sulfur was emitted to the atmosphere. Reconstruction in the case of NOX is slightly more complicated, but basically similar [ibid]. More complex reconstructions of historical emissions have been undertaken recently, e.g. for the Hudson-Raritan estuary [Ayres et al 85].

Conclusions

Despite its length, this paper can be reduced to six relatively simple propositions.

1) Conservation of mass (a consequence of conservation of mass-energy) implies that materials extracted from the earth's crust must eventually return thence as waste residuals. The quantities involved are currently very large, and may well be sufficient to trigger irreversible changes in the earth's atmosphere.
2) The non-existence of any mechanism for pricing the environmental absorptive capacity (a common-property resource) needed to dispose of the waste residuals resulting from the use of fossil fuels and toxic heavy metals, has led to a massive underpricing of those resources, and a correspondingly excessive rate of use. The magnitude of the distortion involved also undermines the plausibility of existing equilibrium-type models.
3) There is no lower limit to the materials required to generate final goods and services (GNP), and the potential for materials conservation has been greatly underestimated.
4) Similarly, there is no lower limit to the energy required to generate final goods and services (GNP) and the potential for energy conservation has been massively underestimated in the past. In fact, the imperative for further economic growth in no way puts any long-term requirements on the amount of energy that will be needed. Even in the short term, energy conservation can easily accommodate any rate of growth that is likely to occur.
5) The fact that our economy is dissipative, and the fact that recycling cannot be 100% efficient (as implied by the second law of thermodynamics) do not imply the impossibility of reaching a sustainable stationary-state in the very long run - at least, as long as the sun shines.
6) The fact that our economy is self-organizing, and that it is currently not in equilibrium with the environment implies that our technology- as reflected by "industrial metabolism" - must and will change fairly radically. Such changes are not accommodated by existing economic

growth models. Hence new approaches to economic growth modelling are needed.

References

Arrow, K.J. and Debreu, G. (1954), Existence of an Equilibrium for a Competitive Economy, Econometrica 22(3).

Ayres, R.U. (1978), Resources, Environment & Economics, John Wiley & Sons, N.Y.

Ayres, R.U. (1982), Coalplex: An Integrated Energy/Resource System Concept, UNEP Seminar on Environmental Aspects of Technology Assessment, United Nations, Geneva, Nov/Dec.

Ayres, R.U., Energy Inefficiency in the U.S. Economy, unpublished manuscript.

Ayres, R.U., Cummings-Saxton, J. and Weinstein, E. (1987), Assessments of Methodologies for Indirect Impact Assessment, Research Report (IRT-468-R/a), International Research & Technology Corporation, Washington DC. (NOTE: prepared for EPA).

Ayres, R.U. et al. (1985), An Historical Reconstruction of Major Pollutant Levels in the Hudson-Raritan Basin (3 Vols), Report by Variflex Corporation for the National Oceanographic & Atmospherican Administration (NOAA), Rockville, MD. Published as NOAA Technical Memorandum, 1988.

Ayres, R.U. and Kneese, A.V. (1969), Production, Consumption & Externalities, American Economic Review, June.

Ayres, R.U. and Narkus-Kramer, M. (1976), An Assessment of Methodologies for Estimating National Energy Efficiency, Winter Meeting, ASME, November.

Boulding, K.E. (1966), Environmental Quality in a Growing Economy, in: Essays from the Sixth RFF Forum, Johns Hopkins University Press, Baltimore, MD.

Bridges, J. (1973), Understanding the National Energy Dilemma (1973), Joint Committee on Atomic Energy, U.S. Congress, Washington DC.

Dasgupta, P. and Heal, G. (1979), Economic Theory and Exhaustible Resources, Cambridge U. Press, Cambridge UK.

Dunkerley, J. (1980), Trends in Energy Use in Industrial Societies, Research Report (R-19), Resources for the Future, Inc. Washington DC.

EPA (1986), National Air Pollution Emission Estimates, 1940-1984, Research Report (EPA-450/4-85-014), Office of Air Quality Planning and Standards, US Environmental Protection Agency, Research Triangle Park, NC, January.

Faber, Malte, Niemes, H. and Stephan, G. (1987), Entropy, Environment & Resources, Springer-Verlag, Berlin.

Georgescu-Roegen, N. (1971), The Entropy Law and the Economic Process, Harvard University Press, Cambridge MA.

Georgescu-Roegen, N. (1977), The Steady State & Ecological Salvation: A Thermodynamic Analysis, Bioscience 27(4), April.

Georgescu-Roegen, N. (1979), Myths About Energy & Matter, Growth & Change

10(1).

Goeller, H. and Weinberg, A. (1976), The Age of Substitutability, Science 191, February.

Gschwandtner, G., Gschwandtner, K.C. and Eldridge, K. (1983), Historic Emissions of Sulfur and Nitrogen Oxides in the U.S. 1900-1980, Research Report (Contract No. 68-02-3511 # 3), Pacific Env. Services Inc., Durham NC, October 5.

Herfindahl, O. (1967), Depletion and Economic Theory, in: M. Gaffney (ed.), Extractive Resources and Taxation, University of Wisconsin Press, Madison WI.

Hoos, I. (1983), Systems Analysis in Public Policy: A Critique, University of California Press. Revised edition.

Kenney, W.F. (1984), Energy Conservation in the Process Industries (Series: Energy, Science & Engineering: Resources, Technology & Management), Academic Press Inc., New York.

Meadows, D. et al. (1972), The Limits to Growth: A Report for the Club of Rome's Project on the Predicament of Mankind, Universe Books.

Murray, F.X. (1976), Energy: A National Issue, Center for Strategic & International Studies, Georgetown University, Washington DC. (NOTE: Update of Bridges 73).

Nicolis, G. and Prigogine, I. (1977), Self-Organization in Non-Equilibrium Systems, Wiley-Interscience, New York.

Organisation for Economic Cooperation and Development (1978), World Energy Outlook, OECD, Paris.

Porat, M. (1977), The Information Economy: Definition and Measurement, Research Report (Special Publication 77-12(i)), Office of Telecommunications Policy, Washington DC, May.

Rosenfeld, A. and Hafemeister, D. (1987), Energy Efficient Buildings, Scientific American.

Ross, M., Socolow, R. et al. (1985), Effective Use of Energy: A Physics Perspective, Research Report, American Physical Society, January.

Schurr, S.H. and Netschert, B.C. (1960), Energy in the American Economy, 1850-1975, Johns Hopkins University Press, Baltimore MD.

Seligman, B.B. (1962), Main Currents in Modern Economics, Free Press, Glencoe IL.

Shurkin, J. (1985), Engines of the Mind, Washington Square Press, New York. Revised edition. (NOTE: Pocket Books).

Solow, R.M. (1974), The Economics of Resources or the Resources of Economics, American Economic Review, 64.

Tilton, J.E. (ed.)(1983), Material Substitution: Lessons from Tin-Using Industries, Resources for the Future Inc., Washington DC.

Tribus, M. and El-Sayed, Y. (1981), The Strategic Use of Thermoeconomic Analysis for Process Improvement, AIchE, August 16-19.

Chapter 7

ECOLOGICAL MODERNIZATION OF INDUSTRIAL SOCIETY - THREE STRATEGIC ELEMENTS

U.E. Simonis

Ecological Structural Change of the Economy

In both the East and the West, economists, planners and engineers are seeking for a solution to the problem of how to change the traditional patterns of resource use. "Perestroika" and "modernization" are two actual catchwords in this process, and new environmental priorities play a part in the envisaged conversion of the economy.[1] Harmonizing ecology and economy in a specific sense relies on the premise that a reduction in the resource input of production (structural change) will lead to an ex ante reduction of emissions and wastes that have a negative impact on the natural environment (ecological structural change).

In order to clarify the relationship between economic structure and environmental impacts, one needs suitable information concerning the material side of production, for environmental protection and resource conservation by the economy - and thus its long-term sustainability - cannot appropriately be described in such terms as income, investments and consumption. One possibility is to select and compare some indicators describing the environmentally relevant features of the prodution process. The availability of environmental indicators such as emission data relating to "representative" pollutants - like sulphur dioxide (SO_2), carbon dioxide (CO_2), and nitrogen oxide (NO_x) - has grown recently (e.g., the Annual Report on the Environment by several industrial nations, by UNEP and the OECD). These indicators concern certain negative environmental effects of production. Less is known on the environmental relevance of the input factors in industrial production, or on the question of which indicators provide environmentally significant information about the structure of the economy. Given the present state of statistics, only a few such indicators can be tested in a cross-national comparison of Eastern and Western countries.

(1) De-linking economic growth from environmentally relevant input factors

Using a set of four indicators (input factors), Jänicke et al. have studied 31 countries of both COMECON and OECD with regard to the relationship between economic structure and environmental impacts.[2] The four factors whose direct and indirect environmental significance is thought to be self-evident were: energy, steel, cement, and freight transport. Regarding their patterns of production and consumption these are environmentally "hard" factors, characteristic of a certain structure of the economy and/or stage of economic development.

The main hypothesis of the research reads like this: Positive environmental effects of structural change of the economy are to be expected by actively de-linking economic growth from the use of environmentally relevant inputs (resources). Such active de-linking would

- result in a decrease of resource depletion and/or environmental pollution;
- mean ex ante instead of ex post (end-of-pipe technology) environmental protection;
- promote those integrated technologies which touch upon several environmental effects (pollutants) at the same time.

Structural change as a shift of input factors to more efficient uses can thus be conceived as a process of successive de-linking: The contribution of traditional (hard) input factors to the national product decreases, i.e., they change or lose their function in the development process.

(2) Examples of marked and deficient de-linking

Taking the Federal Republic of Germany as an example, Figure 1 illustrates a five-fold de-linking from the growth of the Gross Domestic Product (GDP): The de-linking of energy and cement consumption and weight of freight transport from the GDP became apparent during the 1970s; regarding steel, the de-linking process began already in the 1960s. In this way, the structural change of the economy generated environmental gratis effects of various kinds:

- The stagnating consumption of primary energy led to a reduction of harmful emissions (pollutants).
- The relative decline in the weight of freight transport indicates that the volume of materials employed was reduced rather than increased.
- The fall in the use of cement represents a direct gratis effect as far as the emissions from cement factories are concerned; this decrease coincided with the trend towards labor-intensive renovation of the

Figure 1: Structural Economic Change in the Federal Republic of Germany, 1960 - 1987 (1960 = 100)

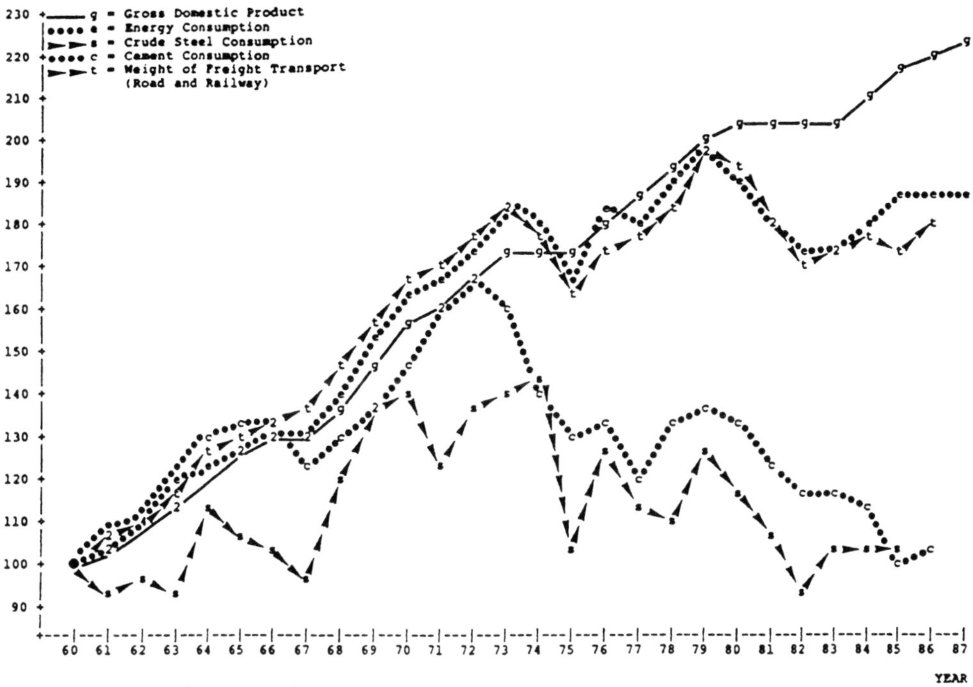

Source: Jänicke et al.

housing stock, as compared to new construction.

- The decrease in steel consumption accounts for a considerable reduction in harmful emissions as far as processing is concerned; this drop was strongly marked and partly due to increased recycling activities.

Environmental gratis effects occur when the rate of usage of the input factors (resources) having a (strong) negative impact on the environment remains below the growth rate of the GDP. Comparing the rates of usage of the four selected input factors with the growth rate of the GDP, three different development patterns emerge[3]:

- The factors having impacts on the environment decline absolutely; i.e., absolute structural improvements are induced, corresponding to absolute environmental gratis effects.
- The factors having impacts on the environment remain constant, or increase, but with a lower growth rate than the GDP; i.e., relative structural improvements, corresponding to relative environmental gratis effects.

- The factors having impacts on the environment increase at a higher growth rate than the GDP; i.e., structural deterioration, corresponding to absolute negative environmental effects of economic growth.

In Table 1, 16 out of the 31 countries studied are grouped according to these three development patterns.

Of all the industrial countries investigated by Jänicke et al., Sweden went through the most rapid structural change. The drastic reduction in cement production (-41%), the decreasing use of crude steel (-38%), and the decrease in the weight of freight transport (-21%) add up to notable environmental gratis effects ("absolute structural improvement").

In Japan, the de-linking process was partly neutralized by the rapid growth in industrial production and thus only resulted in "relative structural improvement".

In Czechoslovakia, no significant de-linking of economic growth from the four input factors took place (see Figure 2). The development profile of this country, with sluggish structural change, is to some extent representative of the other economies of Eastern Europe.

(3) Trends towards industrial restructuring

Despite certain analytical limitations of such empirical research (as e.g., the selection of only four input factors), several conclusions can be drawn from this international comparison as regards to the trends of industrial restructuring[4]:

- Structural change in the form of de-linking economic growth from environmentally relevant inputs was evident in most but not all of the countries studied.
- Several countries enjoyed environmental gratis effects as a result of active structural change. In some cases, especially for Sweden, these effects were quite considerable.
- In other countries, the possibly beneficial environmental effects of structural change were levelled off by the rapid industrial growth pursued. This was especially true for Japan.
- The strong correlation between the level of production (GDP) and environmental impacts, still evident in the 1970s, had dissolved in the 1980s. The high-income countries featured fairly rapid structural change.
- In the medium-income countries, a distinct pattern emerged in that

Table 1: Environmental Impacts deriving from Structural Change - Percentage Changes 1970/1985

Country	Consumption of Primary Energy	Consumption of Crude Steel	Cement Production	Weight of Freight Transport	GDP *
Group 1: Absolute Structural Improvement					
Belgium	7.1	-24.5	-17.6	-2.2	42.7
Denmark	-2.7	-15.6	-33.2	20.1	40.8
France	30.3	-34.8	-23.4	-14.5	51.6
FRG	13.4	-26.3	-32.8	4.4	38.4
Sweden	26.4	-37.9	-41.2	-21.4	32.7
United Kingdom	-2.3	-43.5	-28.7	-18.2	32.4
Group 2: Relative Structural Improvement					
Austria	32.1	-33.9	-6.0	21.3	54.3
Finland	39.6	14.8	-11.2	12.2	65.7
Japan	37.3	-2.3	27.4	7.5	90.2
Norway	51.1	-21.6	-40.3	34.7	87.5
Group 3: Structural Deterioration					
Bulgaria	74.9	24.9	42.3	77.5	37.3
Czechoslovakia	31.5	22.5	37.3	62.9	33.9
Greece **)	119.3	67.3	162.9	43.1	69.1
Portugal **)	89.0	34.2	133.1	27.4	69.0
Soviet Union	76.3	33.4	35.9	70.2	47.7
Turkey	218.8	184.4	173.2	118.6	118.2

*) Calculation of the Gross Domestic Product percentage changes on the basis of constant (1980) US-Dollars. Bulgaria, Czechoslovakia and Soviet Union data refer to percentage changes 1970/83 of the Gross National Product.

**) Transport data only take railway transport data into account.

Source: Jänicke et al.

Figure 2: Structural Economic Change in the CSSR, 1970 - 1985 (1970 = 100)

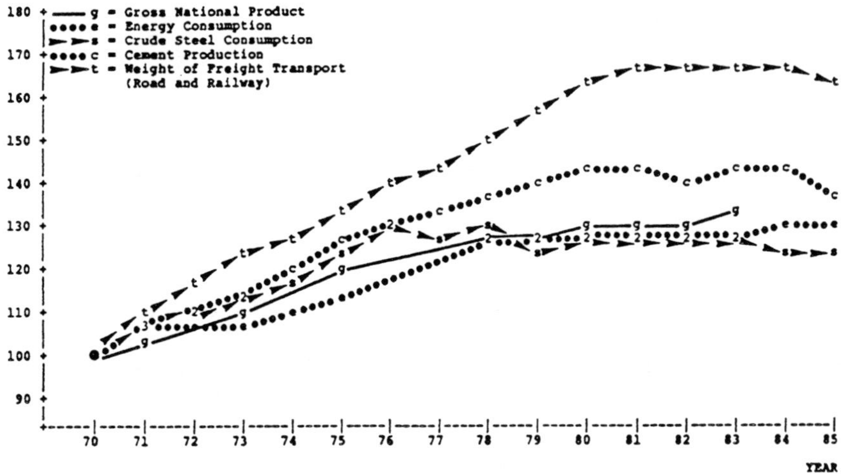

Source: Jänicke et al.

there were cases of rapid quantitative growth and cases of qualitative growth, i.e., economic growth with constant or decreasing resource input.

All in all, it is therefore not yet possible to speak of one dominant trend towards industrial restructuring. However, the environmental gratis effects of active structural change are highly evident and thus provide one strategic element of ecological modernization of industrial society.

Preventive Environmental Policy

Theoretically speaking, environmental policy could be defined as "... the sum of objectives and measures designed to regulate society's interaction with the environment as a natural system; it comprises aspects of rehabilitation, conservation, and structural adjustment".[5] Practice, however, does not conform to such a broad definition. Only parts of the interaction between society and environment become the subject of policy. So far, environmental policy has mostly been designed as react-and-cure strategies concerning air and water pollution, noise, and waste, with emphasis on the rehabilitation aspect.

For a variety of reasons, this conventional environmental policy was and still is meaningful and very much necessary. It has, however, a number of deficits, some of which are cited in the following, along with some suggestions for overcoming them through preventive environmental policy, i.e., anticipate-and-prevent strategies.

(1) Environmental expenditures - environmental damages

Since the beginning of the 1970s, when systematic records first began to keep track of the funds allocated for environmental protection, in the industrialized countries the sum of the respective public and private investments has reached large proportions. In the Federal Republic of Germany, for instance, this sum has come to the handsome total of over 250 billion Deutschmarks (or about 140 billion US-dollars). In a detailed study, Leipert et al. have computed and classified all existing data on investments and expenditures aimed at repairing and protecting the environment.[6] Table 2 shows the total and sectoral environmental protection investments for the manufacturing sector of the German economy for the years 1975 to 1985. Table 3 shows the total costs of environmental protection (current expenditures and depreciations) for both industry and government for the years 1975 to 1985.

Figures like these, however, are ambivalent. On the one hand, they give cause for proud political statements about the successes of environmental protection, according to the motto "the more, the better". On the other hand, they are - presumably - the absolute minimum of what is necessary to secure the very basis for society's sustainability. At the same time, they symbolize a serious structural deficit of industrial society: Environmental protection expenditures are spent when damage to the natural environment has occurred and can no longer be denied. Belated, they are repairs to the process of economic growth, signs of a "post-fact" policy that reacts to damages (and must react to them) but does not, or cannot, prevent them. Therefore, it is necessary to confront the "success stories" of environmental protection expenditures with figures on the environmental damages themselves.

Again taking the Federal Republic of Germany as an example, a recent estimation by Lutz Wicke showed that the annual damage to the natural environment is above 103 billion DM, or in the order of 6% of the GNP, and not 3%, as the OECD had estimated for the industrial countries some years ago.[7]

Table 4 is based on data on actual damage costs and findings from

Table 2: Environmental Protection Investments, Manufacturing Sector, Federal Republic of Germany, 1975 - 1985

Year	Total Investments		Waste Disposal		Water Pollution Control		Noise Abatement		Air Pollution Control	
	Current Prices	1980 Prices	Current Prices	1980 Prices	Current Prices	1980 Prices	Current Prices	1980 Prices	Current Prices	1980 Prices
	in Millions of DM									
1975	2,480	3,090	170	210	900	1,110	200	240	1,210	1,530
1976	2,390	2,830	200	230	820	960	220	260	1,150	1,380
1977	2,250	2,560	200	230	740	850	210	230	1,100	1,250
1978	2,150	2,370	170	180	680	750	200	220	1,100	1,220
1979	2,080	2,190	260	160	760	800	200	210	960	1,020
1980	2,650	2,650	210	210	910	910	240	240	1,290	1,290
1981	2,940	2,810	250	240	950	910	210	200	1,530	1,460
1982	3,560	3,250	390	360	1,130	1,030	230	210	1,810	1,650
1983	3,690	3,270	290	260	1,100	990	230	200	2,070	1,820
1984	3,500	3,100	270	240	1,040	920	230	190	1,960	1,750
1985	5,620	4,940	330	280	1,060	910	260	220	3,970	3,530
	Average Annual Change in %									
1975/84	+ 3.9	- 0.0	+ 5.3	+ 1.5	+ 1.6	- 2.1	+ 1.6	- 2.6	+ 5.5	+ 1.5
1975/79	- 4.3	- 8.2	- 1.5	- 6.6	- 4.1	- 7.9	0.0	- 3.3	- 5.6	- 9.6
1979/84	+ 11.0	+ 7.2	+ 11.0	+ 8.4	+ 6.5	+ 2.8	+ 2.8	- 2.0	+ 15.3	+ 11.4

Source: IIES research project.

Table 3: Total Costs of Environmental Protection, Federal Republic of Germany, in Millions of DM, 1975 - 1985

Year	Industry			Government			Industry and Government		
	Current expen-ditures	Depreci-ations	Total costs	Current expen-ditures	Depreci-ations	Total costs	Current expen-ditures	Depreci-ations	Total costs
	At current prices								
1975	3,200	1,520	4,720	3,000	1,920	4,920	6,200	3,440	9,640
1980	5,160	2,250	7,410	4,690	3,390	8,080	9,850	5,640	15,490
1985	7,930	3,160	11,090	6,430	4,340	10,770	14,360	7,500	21,860
	At 1980 prices								
1975	4,050	1,870	5,920	3,790	2,570	6,360	7,840	4,440	12,280
1980	5,160	2,250	7,410	4,690	3,390	8,080	9,850	5,640	15,490
1985	6,230	2,640	8,870	5,340	4,030	9,370	11,570	6,670	18,240

Source: IIES research project.

Table 4: Environmental Damage in the Federal Republic of Germany ("Measurable Damage" in Billions of DM per year)

Environmental Sectors	Environmental Damage
Air Pollution	ca. 48.0
Health hazards	between 2.3 - 5.8
Material damage	more than 2.3
Degradation of vegetation	more than 1.0
Forest blight, etc.	between 5.5 - 8.8
Water Pollution	far more than 17.6
Damage to rivers and lakes	more than 14.3
Damage to the North Sea and Baltic Sea	far more than 0.3
Contamination of ground water, etc.	more than 3.0
Soil Contamination	far more than 5.2
Costs of Chernobyl disaster	more than 2.4
Rehabilitation of "yesterday's waste"	more than 1.7
Costs of preserving biotopes and species	more than 1.0
Other soil contamination, etc.	far more than 0.1
Noise	more than 32.7
Degradation of residential amenities	more than 29.3
Productivity losses	more than 3.0
"Noise rents," etc.	more than 0.4
Grand total of damage	more than 103.5

Source: Wicke.

willingness-to-pay studies. Though the results must be taken with some care, the table illustrates that despite high annual environmental protection expenditures still enormously high environmental damages occur annually. Of course, this situation may not only be true for Germany but for many other countries as well.

There are more shortcomings of conventional environmental policy.[8] To name a few: Environmental policy usually identifies the given problem too late, so that the ecosystems affected cannot survive. As it is pursued as a media-specific policy, i.e., separately regulating air and water quality, noise or waste, it also runs the risk of lacking coordination between its specific goals, measures and institutions. And this then may result in shifting a problem from one environmental medium to another, e.g., from air

to water or soil, or from one place to another, as is the case with long-range, trans-boundary pollution. In addition, environmental policy often becomes entangled in a debate on principles. If measures must quickly be taken, the argument gets shifted from the "polluter-pays-principle" - which is advocated in general - to the "taxpayer-pays-principle", thus switching the burden of environmental protection from the individual polluter to the community, to government or to society at large.
Thus, innovations in planning and implementation are needed. Preventive environmental policy - it seems - can counter the shortcomings of conventional environmental policy. But in order to switch to preventive policy, several conceptual as well as practical constraints have to be overcome.[9]

(2) Prerequisites of preventive environmental policy

According to Scimemi and Winsemius one can conceive three factors as concomitant policy relevant processes in time: The accumulation of environmental damage; the acquisition of technical knowledge; and the rise of public awareness. The time sequence of these processes, especially the relative timing of their critical level, is decisive for the whole issue of preventive environmental policy.

To illustrate the relationship between these three factors, Scimemi has redrawn a diagram suggested by Winsemius, using three separate functions: Level of Damage, Level of Technical Knowledge, Level of Public Awareness.[10] The relative position and the shape of these functions depends, of course, on the specific circumstances (country, environmental sector, historical phase) under consideration; the common case is illustrated in Figure 3.

Line D_1 - D_2 indicates the accumulation of environmental damage over time. The accumulation of damage starts at a given point in history (in the diagram somewhere between time 0 and time 10). At that point, neither the scientific community nor the general public is yet aware that anything of importance is happening. Line T_1 - T_2 indicates the process of gathering technical knowledge. This process may not start until some time after damage has begun to accumulate (in the diagram somewhere between time 10 and 20), and proceeds gradually. During that phase the public may still be unaware of the hazard. Somewhere between time 20 and 30, public awareness starts to rise, as indicated by the line P_1 - P_2.

Within these concomitant processes, a certain stage becomes important (critical level). As illustrated, the technical understanding of the issue reaches a critical level, t, thus ensuring the first of two conditions

Figure 3: Factors of the Environmental Policy Cycle: Damage, Technical Knowledge, Public Awareness

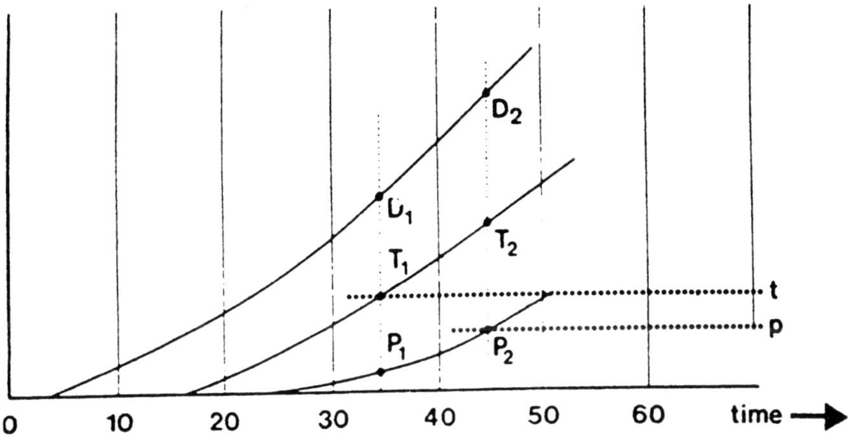

Source: Scimemi.

required for effective policy action, i.e., technical rationality. Public awareness also reaches a critical level, p; at that time the second condition for effective decision-making, i.e., political viability, is fulfilled. It is at this stage that action will be undertaken to avoid the occurrence of further damage.

Recalling past developments in environmental policy at the national or the international level, it is easy to recognize that the processes evolved very much in conformity with Scimemi's theoretical interpretation. Thus, the diagram may be considered to be a true representation of real events. (There are cases, however, where public awareness - line P_1 - P_2 - started much earlier than technical knowledge - line T_1 - T_2).

What are now the opportunities to influence these basic conditions of policy action in favor of preventive environmental policy? Figure 3 helps to formulate three general and two specific options[11]:

General options:

- Retarding damage accumulation (i.e., sliding the D_1 - D_2 curve towards the right).
- Accelerating technical knowledge (i.e., sliding the T_1 - T_2 curve towards the left and/or raising its slope).
- Increasing public awareness (i.e., sliding the P_1 - P_2 curve towards

the left and/or raising its slope).

Specific options:

- Dynamic environmental standard setting (i.e., lowering threshold t).
- Dynamic public participation (i.e., lowering threshold p).

All these various options make policy decisions possible at a stage when the level of environmental damage is still relatively low, or even zero.[12] One of these options is illustrated in Figure 4.

(3) Environmental impact assessment

Acceleration of knowledge and awareness can, of course, be promoted through a variety of approaches and methods and depends a great deal on the specific environmental issue at hand. Environmental Impact Assessments (EIA) are increasingly being applied, not only for public but also for private investment projects. They entail efforts to learn more about possible environmental impacts, and are intended to allow appropriate action to be taken before damage has occurred. In that sense, environmental impact assessments can be classified as part and parcel of preventive environmental policy.[13]

During the last years, some headway has been made to institutionalize and standardize EIA procedures, nationally and to a minor extent also internationally. As the EIA procedure is particularly used for specific investment projects, it allows for the "accelerating effort" to be targeted, and generally also permits the burden of such efforts to be imposed upon the project initiator himself, thus conforming to the polluter-pays-principle. A big deficit, however, remains in how to implement EIA as a preventive procedure in cases of global change, like climate warming or ocean pollution.

The required levels (thresholds) regarding technical knowledge and/or public participation in environmental decision-making differ widely from one environmental medium and country to the other.[14] The question of how much knowledge/awareness is enough normally falls upon the political decision-maker (the government, the environmental protection agency, the institution in charge of the problem), even if the scientific community (or parts of it) is ready to say "we know enough" and the public (or parts of it) is demanding "something must be done". Therefore, stalmates in decision-making on environmental issues are quite frequent.

What is "enough knowledge/awareness" for one country (government,

Figure 4: Accelerating the Generation of Technical Knowledge and/or the Development of Public Awareness

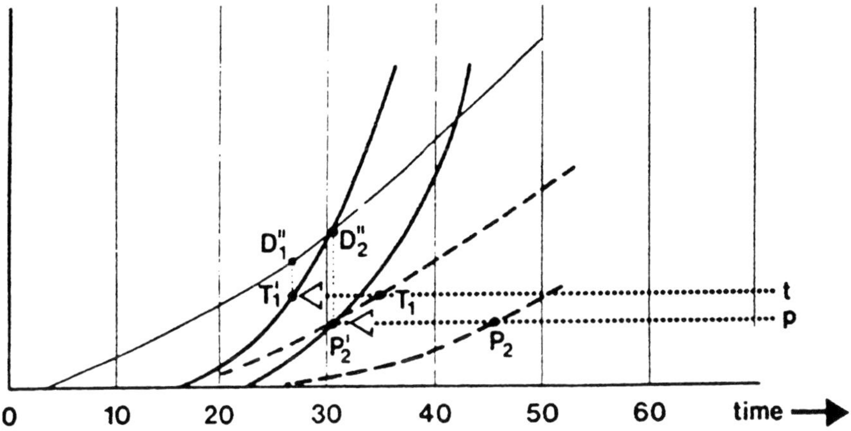

Source: Scimemi.

agency, institution) may not be enough for the other. The normal outcome of such a situation is a meagre compromise over the emission standards to be implemented. They will be weaker than technically/politically feasible because knowledge/awareness on cause-effect-relationships or social priorities is said to be insufficient. Eminent cases in point are the emission standards for SO_2, CO_2 and NO_x in the air pollution field, and the nitrate standard in the water pollution field. Thus, the dilemma of setting stricter emission standards is serious. Meanwhile the forests may continue to die back, the ozon layer may continue to be affected, and the water may continue to get contaminated.

The conclusion therefore is, that environmental standard setting must be conceived as a continuous process. With growing knowledge/awareness on actual and probable environmental damages the thresholds for action must be consecutively lowered, i.e., standard setting must be dynamized to speed up industrial restructuring.

This need to come to terms with the future is, of course, not unique to environmental policy. Implementing the prevention principle is especially requested in all other domains of policy where collective interests are at stake. One such major domain we have to address when discussing the possiblities and impediments of ecological modernization is economic policy.

Ecological Orientation of Economic Policy

"Ecology in essence means the necessary and feasible harmony between man and nature, society and environment." (C.F. von Weizsäcker). Economy, however, in general means disharmony with nature. Use is made of nature both directly and indirectly when raw material are processed into products, and nature is polluted by the emissions and wastes generated by industrial production. These are, then, the two processes in which nature remains the loser. She exchanges natural raw materials for produced waste materials. Besides labor and capital, nature is the truly quiescent and exploited third production factor.

(1) Conflicts between economy and ecology

The use of raw materials and the generation of emissions and wastes are of course old, not new issues. Scientific and technological development, however, has made it possible to increasingly exploit the depletable resources, and has lead to an ever increasing accumulation of harmful emissions and non-decomposable wastes. Nature is no longer able to absorb all of these substances, many of which are not only toxic for flora and fauna but for human beings as well.

Efforts to hide harmful emissions and toxic wastes - in land-fill sites, in transfer stations or permanent depositions, to spread them- through high smokestacks and incinerators, or to dump them - into the water bodies and abroad, have at best been temporarily successful because many emissions and wastes are "mobile poisons" or reappear in different form. These activities lead to a "linearization of ecological cycles,", i.e., the natural diversity is reduced, the robustness of ecosystems declines, ecological symbioses and equilibria break down. As a consequence, environmental degradation increases and the absorption capacity of the natural environment decreases.

Accordingly, the conflict between ecology and economy can be attributed to two (actually or possibly) incompatible basic principles: The ecological principle of "stability," as a precondition for the sustainability of ecological systems, and the economic principle of "growth," as the inherent logic of economic systems - more precisely: the principles of business profitability, national economic growth, and world market expansion.

Given the actual and the pending ecological crisis, the question on whether these economic principles can be changed, reshaped and finally brought into harmony with ecological principles, on which level, in what way, and at what time, is, of course, a controversial question in both theory and practice, and a specific challenge to the social sciences. The answer not only depends upon the respective individual and societal constellation of interests. It particularly depends upon the ability of and the willingness for social innovations, i.e., on (a) whether the potential of an ecological self-regulation of the economy is used, and (b) how the option of an ecological re-orientation of economic policy is implemented.[15]

(2) Ecological self-regulation of the economy?

To start with a general statement: Most certainly, only a small fraction of the current environmental problems would exist if the economic contexts would have remained so comprehensible, that producers and consumers would personally be able, and liable to recognize and perceive the consequences of their own decisions towards depleting resources and polluting nature. Or, if business profitability, national economic growth, and the expansion on world markets could not be increased by externalizing parts of the ensuing costs. This is the old but still relevant - because unresolved - problem of the external effects of production. Scientific and technological development has been, and still is, coupled with negative external effects, i.e., the shifting of costs to society, future generations, and nature. With respect to the environmental problem, all these cost components are relevant.

One of the pending tasks thus can easily be prescribed: Internalize the external effects of production, shift the costs back to the economic units that cause the environmental problem, include the ecological perspective into all investment decision-making. Drastically reducing the external effects of production on society, nature, and future generations seems to be the necessary step towards regaining harmony between economy and ecology. But, how to proceed in practice and where to put priority?

To re-organize the economy towards a materially integrated cycle would, first of all, mean to reduce systematically the use of depletable resources and the generation of polluting emissions and wastes - and this is in contradiction to the prevailing "throughput economy" (K. Boulding). In practice, recycling and clean technology are still at an incipient stage and not systematic economic undertakings. Especially, the step from simply disposing wastes towards avoiding wastes ("low waste economy") has not been made.

Certainly, this is in part because many waste products cannot be recycled or only at high cost. But it is also true because adequate price and cost signals have not been set. Preventing waste generation and conserving depletable resources are still not sufficiently being promoted. This state of affairs, however, has also to do with the deficits of our economic accounting procedures which do not adequately measure the diminishing stocks. Therefore, two contradictory trends can be observed: increasingly monetary income - decreasing natural stock.

Proposals for ecological accounting at the factory level and in the national accounts, however, are promising. With ecological accounting the amount of energy, materials, wastes, land use, etc. are being computed and, by simulating the given shortage, accounting units are determined which then enter the accounts. Thus a measure is developed which not only may guide private investment decision-making, but at the same time will provide a public information instrument for promoting qualitative economic processes.

In industrial society another ecological principle is no longer valid, i.e., sustainability of resource use. Traditionally, forest owners for instance have followed the rule "Do not cut down more wood than you regrow." This rule is being undermined: externally produced acid rain collides with internal resource conservation; and accumulated external debt lead to over-exploitation of national resources. Sustaining the yield of the forest capital stock is being replaced by indirect expropriation or resource depletion.

Another basic principle to be re-established in the economy is that of responsiblity or liability. With respect to environmental problems, the legal system, and also economic behavior, in most countries is marked by the strict proof of causality. Only when the injured (damaged party) can prove who caused the damage (polluting party), the polluter is held liable for compensation. Instead, in some countries (for some cases) statistical probability is sufficient for obligating industry to compensate for damages (collective liability). Once this principle was established by the courts and through legislation, it quickly helped to improve environmental quality through ecological self-regulation of business activities.

In general, the liability principle would support the anticipate-and-prevent strategy in environmental policy, and shift the technical solutions for environmental problems from ex-post to ex-ante approaches, i.e., from controlling or end-of-pipe technology towards low emission or integrated technology. To implement the principle in practice, small steps or big leaps can be taken: from continuous reporting on wastes or automatic monitoring of emissions, to collective funds and to strict environmental liability.

(3) Ecological re-orientation of economic policy?

Confronted with serious environmental damage, conventional econmic policy is increasingly being challenged. Its guiding principles, goals, instruments, and institutions are being questioned, and a new concept is emerging: ecological economic policy.

Conventional economic policy is based on the guiding principle of maximizing flows: volume of production, income, profits, turnover ... (seventeen years ago Kenneth Boulding called this the "throughput economy.") Instead, the "ecological economy" is based on a different guiding principle, i.e., "increasing efficiency while maintaining substance!" Aspects such as environmental compatibility and resource conservation become important, and structural adjustment of products and technologies according to ecological considerations becomes the task.

Regarding goals, it seems necessary to redefine and supplement the conventional economic policy goals, especially to re-assess the growth target and to include "environmental stability" into the catalogue of economic policy goals. The conventional policy goal indicators were developed at a time when environmental pollution was already a problem but not yet a public issue, and since then they have not really been readjusted. Economic growth is still being measured in terms of goods and income categories only (GNP - Gross National Product), while their effects on the stock and the quality of the resources (natural capital) are not adequately considered. In the conventional concept of economic growth, all monetary transactions are summed up independent of their specific function; also, increasingly more expenditures are included which are solely being spent for damage originally caused by the production process ("defensive expenditures").

Qualified goal indicators for economic policy can be defined in various ways: Computations of defensive expenditures, i.e., assessment of an environmentally related net product (ENP - Eco National Product); combined growth and distribution indices (RWG - Redistribution With Growth); an integrated system of economic and environmental indicators, or an attached "Satellite System", etc.

Regarding instruments, conventional economic policy relies mainly on two instruments, i.e., variations of interest rates and of tax rates. From an ecological point of view, new taxes and charges are required which, to some extent, would replace traditional taxes. In a situation of structural unemployment and environmental pollution the introduction of resource taxes (as e.g., an energy tax) and emission charges (as e.g., charges on SO_2, NO_x, and CO_2 emissions), and a definite decrease of wage taxes are called for. Such a structural tax reform would change the existing incentives in the economy towards promoting resource efficiency and increasing employment opportunities.

Economic policy manifests itself in and works through particular institutions. Therefore, an ecological orientation of economic policy also requires creating new institutions, and abolishing or redefining old ones. (The current debate on the negative environmental effects of decisions by the World Bank and the IMF is just a case in point.) The actual and the pending environmental crisis require structural institutional reforms by which economic institutions would have to incorporate the ecological perspective, environmental institutions would have to improve their competence - and by which environmental impact assessments (EIA) would be integrated into all major economic decision-making.

References

[1] Cf. G. Enyedi et al. (1987), Environmental Policies in East and West, London: Taylor Graham.

[2] The following data and arguments rely on M. Jänicke, H. Mönch, T. Ranneberg, U.E. Simonis (1989), Economic Structure and Environmental Impacts, in: Environmental Monitoring and Assessment.

[3] Cf. Jänicke et al.

[4] Cf. Jänicke et al.

[5] International Institute for Environment and Society (1982), Research Program 1983-1987, Berlin, IIES, p. 6.

[6] The following data are from C. Leipert, U.E. Simonis (1988), Environmental Damage -Environmental Protection. Empirical Evidence on the Federal Republic of Germany, in: International Journal of Social Economics, Vol. 15, 7.

[7] L. Wicke (1986), Die ökologischen Milliarden (The Ecological Billions), Munich: Kösel, p. 123.

[8] U.E. Simonis (1985), Preventive Environmental Policy. Prerequisites, Trends and Prospects, in: Ekistics, 313, July/August, pp. 368-372.

[9] In the following I rely on the arguments put forward by G. Scimemi (1988), Environmental Policies and Anticipatory Strategies, in: U.E. Simonis (ed), Präventive Umweltpolitik, Frankfurt, New York: Campus.

[10] Cf. Scimemi (1988).

[11] Cf. Scimemi (1988).

[12] For a detailed illustration cf. Scimemi.

[13] Cf. P. Wathern (ed.) (1988), Environmental Impact Assessment. Theory and Practice, London: Unwin Hyman.

[14] Cf. Scimemi (1988).

[15] Cf. U.E. Simonis (1988), Ecology and Economic Policy, in: A. Vlavianos-Arvanitis (ed.), Biopolitics, Athens: Biopolitics International Organisation.

Chapter 8

ECONOMICS VERSUS ECOLOGY

S. Lombardini

Economic Systems and Natural Systems

An economic system is usually considered as an open system. Inputs come both from the natural and the social-cultural system. The natural system provides the **natural resources,** in the economist's sense, and all those resources and conditions which, even if they do not have any economic value, are required for production activities or are useful to consumers. The inputs from the social system have been summarized by Pareto with the term **tastes** (gusti). In fact the social system provides other inputs viz. those social factors which contribute to mould the attitude to business (propensity to entrepreneurial activity) and to shape the relationship between competition and cooperation.

The usual concept of an economic system (as an open system) does hardly allow us to consider the feed-back effects of its working on both the natural and the social system. We are, however, not concerned here with the latter effects. We shall make a few remarks on the former effects that can be divided into two groups:

a) changes in the natural system that are meant to improve the quality of the inputs for the economic system, e.g. land improvements (positive effects)

b) changes that cause a deterioration of the natural system, such as pollution (negative effects)

It is difficult to draw a line between the two classes of the economic feedbacks on the natural system. Some effects that in the short term look as positive effects may result as negative in a longer term. Such are the effects of some improvements of land for agriculture.

The term mostly used by economists to deal with the feedback effects on the natural system is **environment.**[1] We shall reserve this term to deal with some global features of the natural system that can be affected by

---

[1] According to Siebert (1981) environment "may be understood to be the set of natural conditions that defines the human living space. It has become customary to distinguish among different environmental systems such as air, water and land" (pp. 7-8)

economic activity and that are relevant both for consumption and for production. **Natural system** is a broader concept, for it leads us to consider both environment and single elements such as land, different strata of atmosphere, water and solar energy. One fundamental feature of the feedback effects on the natural system are the interdependencies that appear among feedback effects both in space and in time. As Kneese (1971, p. 16) reminds us "high stacks on power plants in England are said to be causing 'acid rain' in Scandinavia". As Siebert (1981) reminds us "Pollutants ambient in the air can be deposited into water systems, and pollutants ambient in rivers, lakes and the oceans can get into the atmosphere by evaporation. Similar relationships exist between air and land as well as water and land. Besides diffusion in a physical sense, for instance through ground water systems, diffusion may occur through ecological systems such as food chains (bio-diffusion)" (p. 145).

Interdependencies in time arise because of the cumulative feature of some feedback effects on the environment (climate) and on single elements of the natural system (water, atmosphere).

We must remember that we do not know yet the global laws of the universe and of our planet with all organisms living on it (cosmic laws). Our knowledge is confined to local laws. We are thus unable to forecast the effects - for instance - that changes in the species of animals brought about by genetic engineering innovations may have on the ecological equilibrium. Such an equilibrium has been disturbed even by the limited effects that have been caused on some species by some agriculture technologies. Environment is relevant both for consumption and for production. It can be defined according to different spatial dimensions. As aspects of a **local environment** we can consider:

a) the environment salubrity, namely the intensity of the effects that the environment and the various elements of the natural system have on health.[2] We must distinguish these effects from those that are produced by the specific ways in which some natural resources are used, for instance by the technologies applied in agriculture. In fact, links usually do exist between the two kinds of effects

b) the environment utility (landscape consumption value), namely those properties of the environment that provide utility to people living in the surroundings

---

[2] There are serious methodological problems in estimating the effects of pollution on health, see Lave and Seskin (1971).

c) the environment productivity, namely those properties that contribute to the productivities of given factors of production in the region (as, for instance, the productivity of water for fishing and for agricultural uses that is the higher the less it is polluted)

d) the environment stability, namely its capability of maintaining some essential features for sufficiently long time

There are some difficulties in properly perceiving the stability property, for the evolution of environment may be characterized by discontinuities (catastrophe points, points at which some evolutionary processes cannot be reversed). Moreover, the stability property cannot be correctly assessed with reference to local environments because of the space interdependencies already mentioned.

Limitations of Neoclassical Economic Theory in Relation to Environmental Problems

In economic theory all relevant activities are either exchange activities or activities that can be assimilated to exchange activities. The ultimate end of exchange activities is the satisfaction of consumers' needs.[3] Therefore, the central problem of economics is the (market) value problem.

Production is reduced to an equilibrium established by technology between inputs that are obtained on the market (and have thus a market value) and outputs that will be sold (or could be sold) on the market. Natural resources are inputs, since being perceived as scarce, they can command a price on the market where they are sold and purchased.

Consumption is essentially the use of commodities by individuals. Such concepts of production and consumption are inadequate since:

1) they cannot give account of innovation activity, as this implies propensities that cannot be reduced to the attitude to exchange commodities in a successful way, by exploiting existing knowledge for a better utilization of available resources. That has been clearly understood by Schumpeter. In fact in neoclassical models technological progress is essentially considered as an exogenous activity

2) by considering individual tastes as given, the theory hinders all possibilities of analyzing the feedbacks of economic activity on the social system

---

[3] The State activities are justified, for they create the conditions for a safe and efficient satisfaction of consumers' needs.

3) to properly understand production and consumption processes, a holistic approach is required. Production entails processes that cannot be assimilated to exchange activities. In fact, the environment has to provide resources (and conditions) that are not marketable. Such conditions establish links among the various production processes. Consumption, too, cannot be understood by establishing a functional relation between the purchase of commodities and some utility index. The meaning that the use of single commodities has for an individual depends upon the environment and the social system in which consumption occurs. The inadequacy of the neoclassical concept of production and consumption are particularly relevant when we deal with ecological problems. To understand such a statement better, we shall consider the methods to deal with such problems in the neoclassical context.

Neoclassical Approaches to Ecological Problems

Neoclassical economists have been aware - at least since Marshall and Pigou - of consequences of production and consumption activities that do not entail costs or revenues for the agents responsible for them. To analyze such consequences the concepts of external economies and diseconomies (externalities) have been proposed.

Pollution can then be considered as an external diseconomy caused by production processes or consumption. Such externalities have been dealt with by some economists[4] as two parties problems. If the emittor of pollution and the receptor are agents that can make specific agreements, then the problem of pollution can be solved **through private exchange**. The essential feature of the neoclassical model can thus be preserved. Unfortunately the cases in which such a condition materializes are rare and irrelevant. In general, the agents damaged by the pollution caused by specific activities are large in number. Most of them are not sufficiently aware of the damage. In any case it is practically impossible to strike a deal between the agents responsible for the pollution and all people who are affected or will eventually be affected.

There is another aspect of pollution that makes the approach we are considering absolutely inadequate. In fact, pollution is the result of various production processes independently carried out: it established some peculiar kinds of complementarities - between different production and consumption activities - that can be analyzed if the required holistic

---

[4] See Buchanan and Stubblebine (1962).

approach is not hindered by the reduction of production to exchange activities and of consumption to individual acquisition of commodities.

A different approach, conceived in the context of neoclassical economics, relies on the enlargement of the set of original resources that should command a price because of their direct or indirect usefulness. Water and air utilized to discharge wastes must be paid for. That can be done in different ways: by a license that must be obtained in order to carry out some activities that cause water and air pollution, by a tax that has to be paid in proportion to the discharge of polluting wastes, by some kinds of rent charge etc.

There are situations in which such a model to deal with pollution problems can be applied. They are essentially those in which pollution, if it is made sufficiently costly, will be kept within limits such that the negative effects on the environment do not accumulate, nor do they interact negatively with the effect of other polluting activities.

Baumol and Oates (1971) have analyzed how, by a proper system of unit taxes and subsidies, the government can achieve a set of specified standards for environment quality. Their model seems more reasonable than others since it presupposes only that producers are always trying to reduce costs. Yet such an assumption may not be valid in some market forms.

An approach more promising than those suggested by the concept of externalities is that based on the concept of **public good** (see Malinvaud, 1971). Such an approach raises a serious problem: how we arrive at the preferences on the basis of which the choices concerning the public goods are made. In the context of neoclassical theory the problem is conceived as an aggregation problem: how social values can be derived from individual preferences. Yet the problem is much more intricate than usually perceived. My colleagues adhering to contractualism have not yet been able to solve what I proposed to them as the **Navona Square paradox**. When the Rome municipality decided to close Navona Square to car traffic, shopkeepers and restaurant owners in the square were all against the decision that had been forced upon them. If Rome municipality would now try to open Navona Square to car traffic, they would have to face the hostility of the same shopkeepers and restaurant owners. Correct decisions concerning public goods can be arrived at only through some kind of interaction between proposals made directly by public administrations aware of possible future perspectives and individual preferences that are dependent on past experiences and information.

Social assessments of the effects of pollution and social determination of ecological goals are required even in those approaches - similar to the Baumol and Oates' ones that have just been referred to - that can be conceived in a context akin to the neoclassical one. This is because of the interdependencies just mentioned and the need to estimate in advance the quality of the environment that we want to restore or to preserve. The new approach that I proposed (Lombardini, 1987) - the **computational economics approach** - may help to solve these problems.

The problems posed by pollution to economic policy are more complex than those that have been dealt with by the essentially static neoclassical models. The pollution effects of some production and consumer activities can in fact be reduced by proper investments. Such investments will reduce consumption, contributing also in this way to reduce pollution. D'Arge (1971) has analyzed the effects on the economy of these policies against pollution by Harrod type models.

Information, Interest and Ethical Values

Whatever solution we can find to the problem of social values and individual preferences, its implementation depends on the information that contributes to shape individual preferences. In fact, also because of the discontinuity of the possible effects of economic activities on the environment, most individuals do not possess adequate information on the damages that will eventually be caused by pollution. On the contrary, there may be individuals assuming a neurotic attitute towards ecological problems. The consequence may be the same in both cases. None of these people are willing to pay what is required to preserve the environment: the former because they do not perceive the damages that can and must be avoided, the latter because they feel that there can be no remedy for the collapse of the planet.

Differences can arise among individuals with regard not only to information available but also to their interest in the control and in the reduction of pollution. Only if the individual awareness of the possible damages of pollution is sufficiently strong and the interest in avoiding them sufficiently great and uniform, we can expect that efficient political decisions aiming at controlling and reducing pollution can be taken and implemented. It is also true that only the expectation of efficient political decisions can induce individuals to support proposals of interventions to preserve the environment that are costly or that limit individual freedom. If such a relation between political proposals and individual expectations cannot be established, there is no procedure that can lead to the solution of environmental problems by decisions justified in the neoclassical framework.

Such a condition is not easy to be obtained, for the reason that the cost of intervention is borne by the present generations and the damage is felt mostly by future generations.

Siebert (1981) who has produced an interesting neoclassical analysis of environmental problems has noticed that "the demand for a long-run thrust of environmental policy contrast markedly with the actual environmental policy observed. Of course, the old landfills are an example of a rather short-run oriented environmental policy not anticipating future damage. The policy maker very often has an extremely high discount rate, and critics say, four months before an election it is well above thirty percent." (pp. 144-145).

In fact, a market economy is not able to assure the preservation (and possibly the improvement) of the natural system beyond the horizon and the levels which are justified by the interests that the present agents can express on the current markets. From this point of view some of the premercantilistic economies were superior. Some of the customs in a primitive economy (and the associated values) were such as to assure a certain degree of social security and of preservation of environment. The social role of the family contributed to produce similar results in a more recent past.

In fact when individuals' interests are not sufficient to guarantee the essential properties of the natural and the social system, their preservation can be obtained only if individuals abide by some ethical norms. Even in capitalist societies, the market can work because the agents accept the ethics of private property and of the contracts. In fact the law system can make some norms enforceable because they are ethically grounded and can thus be commonly accepted.

After the remarks we have made, it is easy to realize that the solution of ecological problems is possible only if new ethical values are efficiently proposed and widely accepted. Such values should be perceived and accepted as **cosmic values**, namely they must be such as to be acceptable to people of different ideologies and religions. Cosmic values must be recognized in order to avoid also some possible consequences of future technological progress (in genetic engineering, in biology and informatics).

Such a need appears more clearly if we consider the difficulties that a country has to face when it tries to solve ecological problems by itself. In the short run the solution of such problems entails increases in the cost of production of a large set of commodities. Should a country be alone in solving ecological problems, it would have to face an increase in the deficit of its balance of trade. The international community could take the decision required to save environment more easily, by establishing import duties that all countries are entitled to impose for those commodities that come from countries not contributing to ecological policies devised first of all because of the ethical values widely accepted, and then on the basis of economic interests conceived in a sufficiently large horizon.

Ecology in the Perspective of Technological Progress and Evolution of Social Systems

The main obstacle to the implementation of an ecological policy - for a proper restoration and preservation of environment - is assumed to be the cost of such a policy to be borne either by producers or by consumers. The effects of this policy on the balance of trade may cause a reduction in Gross National Product and in employment.

It is not difficult to understand why the analysis of the problem in the context of the neoclassical theory may lead to an overestimation of the cost. In fact taxes and constraints imposed on consumers to implement an environmental policy may induce technical progress capable of reducing pollution. A case in point is what happened in a small lake in Italy some twenty years ago. The municipality forbade a chemical company to discharge liquid polluting wastes into the lake. The firm carried out technical research in order to find a way to solidify the wastes. As a result of their research a by-product was found that could pay for the new plant required to abide by the municipality's order. The new non-polluting technology was not more costly than the old polluting one. Before the constraint was imposed, technical progress was geared to different aims: pollution was not a problem for the firm.

At this point it may be worthwhile mentioning a prejudice of most classical and neoclassical economists. According to them technical progress is predetermined in its potentialities. Institutions as well as entrepreneurial activities can accelerate the progress; they cannot substantially change its direction. Such a prejudice should be dismissed. Technical innovations pursue, more or less consciously, the economic and social goals that private and public agents contributing to them want to achieve. The cost of environmental policy can be reduced also as a consequence of a change in the structure of final demand. The change can occur because of changes in relative prices and of changes in cultural values.

The system of taxes and constraints by which an ecological policy can be implemented will cause increases in the relative values of those commodities the production of which entails a higher degree of pollution. That will lead to a reduction in the demand for these commodities (and therefore in their production) and to an increase in the production of less polluting commodities.

The structure of final demand (consumption standards) has been brought about by the consumer revolution characterized by the sales promotion activities aiming at changing consumers' tastes. It is not easy to assess the rationality of the system that has emerged from such a revolution[5], nor is it possible to analyze all its implication for the environment (see Kneese, 1971). What we can say is that such a system cannot be established all over the world. We cannot hope that life will survive on our planet if all Chinese, Indians, Latin Americans and Africans will live as North Americans do now. Fortunately, the information revolution makes new consumption standards that entail reduction in pollutions possible. These new ways of life will become possible if changes are produced in the social-cultural system. The changes which were required to make the consumer revolution possible were induced by the activities of the firm we have already mentioned. The changes which will open the way to a new informatic society can materialize only if proper formation and information activities are carried out by public bodies.

---

[5] It is not unlikely that in the eye of future historians, the economic development of this century will look more like trumph of irrational greed fostered by producers to increase their profits than as a rational process of development capable of creating more space for rational development of human beings.

References

D'Arge, R.C. (1971), Essay on Economic Growth and Environmental Quality, in: The Economics of Environment (P. Bohm and A. Kneese, eds.), Macmillan, London.

Baumol, W.J. and W.E. Oates (1971), The Use of Standards and Prices for Protection of the Environment, in: The Economics of Environment op. cit. (see d'Arge).

Buchanan, J.M. and W.C. Btubblebine (1962), Externality, Economica, November.

Kneese, A.V. (1971), Background for the Economic Analysis of Environment Pollution, in: The Economics of Environment op. cit. (see d'Arge).

Lave, L. and E.P. Seskin (1971), Health and Air Pollution, in: The Economics of Environment op. cit. (see d'Arge).

Lombardini, S. (1987), Prolegomena to a Theory of Economic Development, Rivista Internazionale di Scienze Economiche e Commerciali, XXXIV, no. 10.

Malinvaud, E. (1971), A Planning Approach to the Public Good Problem, in: The Economics of Environment op. cit. (see d'Arge).

Siebert, H., (1987), Economics of the Environment, Theory and Policy, Springer-Verlag, Berlin.

PART B

ENVIRONMENTAL ASSESSMENT AND POLICY EVALUATION

Chapter 9

AN ASSESSMENT OF ENVIRONMENTAL PROGRESS: THE REASON FOR FAILURE

B. Commoner

Introduction

The profound and troublesome relation between environmental quality and economic development has been a major topic of international discussion beginning with the Stockholm Conference in 1972 and continuing through the Brundtland report. Numerous studies, conferences and commissions have tried to explain to the world that these urgent goals are not in conflict and that both can be achieved. It is time now, after 16 years of effort, to ask a very serious, potentially embarrassing question: What progress has been made in solving either part of the problem; do we understand how the two parts are linked?

I am afraid that the answer is indeed embarrassing. Evidence that neither the environmental crisis nor its disastrous linkage to the economic development of developing countries has been resolved is concisely summarized in a telling statement in the Brundtland report:

"...The industries most heavily reliant on environmental resources and most heavily polluting are growing most rapidly in the developing world, where there is more urgency for growth and less capacity to minimize damaging side effects." (Brundtland, 1987)

My purpose here is to show, from the relevant data, that despite considerable effort, developed countries have largely failed to improve the heavily polluted environment; that the reason for this failure also accounts for the still-unresolved conflict between environmental quality and economic development; and that the remedy for both lies in understanding the origin of the environmental crisis.

With the benefit of hindsight, it is apparent that the failure was predictable at Stockholm. While the Conference generated a widespread commitment to action, it failed to even consider the origin of the environmental crisis. Cures were proposed in the absence of a diagnosis.

Two contradictory theories about the origin of the environmental crisis have been put forward. One still popular idea is that pollution originates from the pressure of increased population and levels of consumption on the ecosystem's inherently limited resources. The contrary view is based on evidence that the escalating levels of pollution in industrialized countries after World War II result from the introduction of ecologically harmful technologies of production rather than from increasing population or levels of consumption (Commoner, 1972a, 1972b). Thus, according to one view, the environmental crisis can be resolved by controlling population and consumption; in contrast, the opposing view emphasizes the need to choose technologies of production that are in harmony with the ecosystem.

At the time of the Stockholm Conference, the data that could be used to resolve this controversy related to the rising levels of pollution in the decades following World War II, when relatively little was done to combat the environmental crisis. Today, with the introduction of corrective measures since the early 1970s, we have a new source of data. We can now ask: which ameliorative methods have effectively reduced pollution, and which have not? Such a comparison can help explain the failures, define the principles of success, and provide new insights into the origin of the environmental crisis and hence its relation to economic development.

Empirical Evidence

In the United States, data are available that provide a fairly detailed picture of the trends in pollution levels since the enactment, in the 1970s of extensive remedial legislation. For example, since 1975 the U.S. Environmental Protection Agency (EPA) has published consistent sets of data regarding the annual emissions and ambient concentrations of a series of standard air pollutants: particulates, sulfur dioxide, nitrogen oxides, carbon monoxide, ozone, volatile organic compounds, and lead. Data on water pollution are less comprehensive, but nevertheless describe the time trends in the concentrations of basic water pollutants such as fecal coliform bacteria, phosphate and nitrate. From these and more scattered reports, it is possible to determine what changes in the levels of a number of environmental pollutants have been brought about by the considerable effort, in the United States, to reduce them (Commoner, 1987a).

Perhaps the most striking aspect of these data is their very wide range. At one extreme are a few clear-cut successes - the few pollutants that have decreased in environmental levels by 70% or more in the last 10-15 years. On the other hand, most pollutant levels have decreased by only 10-20% in that period, and some have actually increased. Given this range of effects, it is possible to relate the magnitude of changes in pollution levels to the types of corrective measure employed and thereby to identify the reasons for the successes and the failures.

The changes in air pollution (Table 1) can be summed up fairly simply. For all the standard air pollutants except lead, the average annual rate of emission has declined only moderately - by 14.1% between 1975 (when consistent measurements began) and 1985. In that period, the annual emission of nitrogen oxides actually **increased** by about 4%. This is hardly the sort of accomplishment envisioned in the environmental legislation adopted in the early 1970s. On the other hand, lead emissions have decreased by 86% (and lead concentrations in blood declined about 40%) in that period, an accomplishment that does approximate the goal of solving the environmental crisis.

There is a similar situation in water pollution. A recent survey of water pollution trends at some 300 sites in U.S. rivers shows that between 1974 and 1981 there has been no improvement in water quality at nearly 90% of the test locations (see Table 2). Concentrations of fecal coliform bacteria, dissolved oxygen, suspended sediments, and phosphorus improved at only 13-17% of the test sites. The nitrate pollution problem has become progressively worse: nitrate concentration increased at 30% of the river sites and decreased at only 7% of them. For the five standard water pollutants, sites with improving levels averaged 13.2% of the tested locations; sites with deteriorating levels averaged 14.7% of the total; 72.1 were unchanged.

If a reduction of 70% or more in national pollution levels is taken as evidence of a qualitative solution of the problem, a search through the available U.S. data reveals that only a very short list of pollutants can meet this criterion: lead, DDT (and related insecticides), PCB, mercury in fresh waters, radioactive fallout from nuclear bomb tests, and in some local situations phosphate. Of course, in certain localities a river or a lake has been greatly improved by halting the dumping of specific pollutants into it. But as the trend data show, there has been little or no overall improvement nationally.

Although they are less complete, European environmental data follow the pattern evident in the U.S. data. For example, between 1978 and 1982, sulfur dioxide emissions decreased by an average of 26% in European countries, while average nitrogen oxide emissions declined by only 1.7% (excluding a 358% increase in Poland). Environmental changes in the Baltic Sea closely resemble those in Lake Erie, where because of continued phosphate and nitrate pollution, eutrophication has persisted and oxygen levels have declined; yet because they have been banned, DDT and PCB in fish have decreased. Similarly, in the Baltic Sea, between 1979 and 1984 average oxygen levels decreased by 11%, phosphate concentrations increased by 101%, and nitrate concentrations by 37%. And, as in Lake Erie, for apparently the

Table 1: Changes of Emissions of Standard Air Pollutants, United States, 1975-1985

Pollutant	Emissions		
	(million metric tons/year) 1975	1985	Percent change
Particulates	10.4	7.3	-29.8
Sulfur dioxide	25.6	20.7	-19.1
Carbon monoxide	81.2	67.5	-19.1
Nitrogen oxides	19.2	20.0	+4.2
Volatile organic compounds	22.8	21.3	-6.6
Average			-14.1
Lead	147.0	21.0	-85.7

Source: U.S. EPA, 1986, 1987

Table 2: Water Quality Trends in U.S. Rivers, 1974-1983

Pollutant	Trends in Concentration (percent of sites)		
	Improving	Deterioration	No change
Fecal coliforms	14.8	5.2	80.0
Dissolved oxygen	17.1	11.1	71.8
Nitrate	7.0	30.3	62.7
Phosphorus	13.1	11.3	75.6
Suspended sediment	14.1	14.7	71.2
Average	13.2	14.7	72.1

Source: R.A. Smith et al., 1987

Table 3: Significant Improvements in U.S. Pollution Levels

Pollutant	Time Period	Percent Change	Control Measure	Reference
Lead emissions[1]	1975-1985	-86	Removed from gasoline	U.S. EPA, 1986, 1987
DDT in body fat[2]	1970-1983	-79	Agricultural use banned	U.S. EPA, 1984a
PCB in body fat[2]	1970-1980	-75[3]	Production banned	U.S. EPA, 1984a
Mercury in lake sediments[2]	1970-1979	-80	Replaced in chlorine production	U.S. EPA 1984b
Strontium 90 in milk[2]	1964-1984	-92	Cessation of atmospheric nuclear tests	U.S. EPA 1984a
Phosphate in Detroit river water[2]	1971-1981	-70	Replaced in detergent formulation	U.S. EPA, 1984b

NOTES:
1. Measured as amount emitted per year.
2. Measured as concentration.
3. Change in percentage of people with PCB body fat levels greater than 3 ppm.

same reasons, the levels of DDT and PCB in fish improved considerably, by 80% and 45% respectively. (U.N. Economic Commission for Europe, 1987).

What can be learned from these observations? Table 3 lists the few environmental successes in the United States and the measures used to achieve them.

There is a common theme in the successful remedial measures. In each case the pollutant was prevented from entering the environment not by using a control device to recapture it after it was produced, but by simply stopping its production or use. Thus, the sharp drop in lead emissions is the result of removing lead from gasoline; the environment is less polluted with lead because less of it is now being used. In the same way, the declining environmental levels of DDT have been achieved because the insecticide has been banned from U.S. agriculture; it is therefore no longer being disseminated into the environment. Similarly, the decline in environmental PCB followed legislation that banned its production and use. There has been a sharp decline in strontium 90 levels because atmospheric nuclear bomb tests, which produce it, have nearly ceased since 1963. In certain rivers, phosphate concentrations have been sharply reduced by banning the use of phosphate-containing detergents; as a result, that much less phosphate is sent down drains, into the aquatic ecosystem. All of these successes have been achieved by a very simple and highly effective method: production of the pollutant is stopped or very greatly decreased.

In contrast to these successes, control measures that are designed to recapture the pollutant after it is produced are ineffective. For example, the control devices that are designed to recapture or destroy air pollutants - such as the power plant scrubbers that trap sulfur dioxide or the catalytic converters that destroy carbon monoxide in automobile exhaust gases - have had little impact on emissions. Thus, the nearly two decades of effort to improve environmental quality teaches us a fairly simple lesson: pollution levels can be reduced enough to at least approach the goal of elimination only if the production or use of the offending substance is halted; the control device strategy is ineffective.

The sharply reduced level of mercury in lake sediments is a particularly informative example. This came about when it was discovered that the major sources of environmental mercury in the Great Lakes were chemical plants that produce chlorine by electrolysing a brine solution (mercury is used to conduct the electric current). Required to give up this practice, the plant operators substituted a semipermeable diaphragm for mercury in the production process. The plants no longer dumped mercury into nerby rivers for the simple reason that they were no longer using it. This was not accomplished by reducing the output of the plant's product, chlorine. On the contrary, as shown in Table 4, during the period 1970 to 1979, when mercury pollution declined sharply, the total national production of chlorine increased by 26%. The mercury pollution problem was solved by **changing the means of producing chlorine**, rather than by producing or consuming less of it.

The same pattern is evident in lead pollution. In this case, the production process is automobile travel, and what is "consumed" is passenger-miles traveled. As shown in Table 4, between 1975 and 1984, while vehicular lead emissions declined by 72%, passenger-miles traveled **increased** by 26%. Clearly, this considerable environmental improvement was not achieved by limiting consumption of the good, but, again, by changing the technological means of production.

DDT provides a similar example. Here the good produced for consumption is the crop that DDT protected from insects, in the United States largely cotton. Between 1970 and 1984, environmental levels of DDT decreased by 70-80%; yet the production of cotton increased by 31%. Again, what was changed was not the amount of good produced or consumed, but the technological means of producing it.

Table 4: Changes in Output of Production Processes with Significantly Reduced U.S. Pollution Levels

Pollutant	Relevant Goods Produced	Time period	Change in Amount of Goods Produced (%)	Reference
Lead emissions	Automobile passenger-miles	1975-1985	+26	US Dept. of Commerce, 1986
DDT in body fat	Cotton	1970-1985	+31	US Dept. of Commerce, 1986
PCB in body	Electrical transformers	1970-1985	-12	US Dept. of Commerce, 1986
Mercury in lake sediments	Chlorine	1970-1979	+26	US Dept. of Commerce, 1985

These examples help to define the meaning of the changes that have brought about the few sharp declines in environmental pollution. Clearly, what has been changed is not population pressure or the level of consumption, but the technology of production.

Technology Changes

Most of our environmental problems are the inevitable result of the sweeping changes in the technology of production that transformed the U.S. economic system after World War II: the new large, high-powered smog-generating cars; the shift from fuel-efficient railroads to gas-guzzling trucks and cars; the substitution of undegradable and hazardous petrochemical products for biodegradable and less toxic natural products; the substitution of fertilizers for manure and crop rotation, and of toxic synthetic pesticides for lady bugs and birds; the introduction of trash-burning incinerators instead of recycling.

Environmental degradation is built into the technical design of these modern instruments of production. A high-compression car engine is not only high-powered, but a smog generator. A farm that uses chemical fertilizers and pesticides is not only highly productive, but also an uncontrollable source of water pollution. A trash-burning incinerator not only produces energy, but dioxin as well. And the environmental hazard is just as much an outcome of the facility's technological design as is its beneficial product. High compression is the cause of both the auto engine's power and its production of nitrogen oxide - which triggers smog. The extensive use of fertilizer and pesticides accounts for the productivity of the modern farm- and for the pollution of rivers and ground water as well. The same combustion process that extracts energy from trash also releases the chemical precursors that then combine to produce dioxin.

By 1970 it was clear that these changes in the technology of production are the **root cause** of modern environmental pollution. Now this conclusion has been confirmed by the sharply divergent results of the effort to clean up the environment. Only in the few instances in which the technology of production has been changed - by eliminating lead from gasoline, mercury from chlorine production, DDT from agriculture, PCB from the electrical industry, and atmospheric nuclear explosions from the military enterprise- has the environment been substantially improved. When production technology remains unchanged, and an attempt is made to trap the pollutant in an appended control device - the automobile's catalyst or the power plant's scrubber - environmental improvement is at best only modest, and in some cases (such as nitrogen oxides) nil. When a pollutant is attacked at the point of origin, it can be eliminated; once it is produced, it is too late. In sum, environmental pollution is a nearly incurable disease; it can only be **prevented**.

Unfortunately, most environmental regulations - certainly in the United States - were created without reference to the origin of the crisis that they were supposed to solve. Because environmental regulation has ignored the origin of the problem, it has dealt only with its subsequent effect. And, having defined the disease as a collection of symptoms, the regulations mandate only palliative measures. The notion of **preventing** pollution - the only measures that really work - appears but fitfully in U.S. environmental laws and has never been given any administrative force.

Instead, regulations establish some "acceptable" level of harm from environmental pollutants (for example, a cancer risk of one in a million) and emissions or ambient concentration standards that can presumably achieve this risk level are established. The net result is that the "acceptable" pollution level is frozen in place. The industries, having heavily invested in equipment designed to just reach the required level, are unlikely to invest more in further improvements. The public, having been told that the accompanying hazard to health is "acceptable", is likely to be equally satisfied. Some optimistically inclined people will look upon exposure at the acceptable level as a kind of guarantee of health. Others, perhaps aware of the linear relation between pollution level and the risk to health, will conclude that we are doing as much as we can and will, in most cases, accept the remaining risk fatalistically.

Clearly this process is the inverse of the preventive, public health approach. It strives not for the continuous improvement of the environment, but for the social acceptance of some, hopefully low, risk to health. In a way this is a return to the medieval approach to disease, when illness - and death itself - was regarded as a debit on life that must be incurred in payment for original sin. Now we have recast this philosophy into a more modern form: some level of pollution and some risk to health is the unavoidable price that must be paid for the material benefits of modern technology.

The preventive approach aims at progressively reducing the risk to health; it does not mandate some socially convenient stopping point. Public health programs did not, after all, decide that they could stop when the risk of smallpox reached one in a million; instead, they persisted until smallpox was eliminated world-wide, bringing the risk to zero. In contrast, the present regulatory approach, by setting a standard of "acceptable" exposure to the pollutant, erects an administrative barrier that blocks further improvement in environmental quality. This is, I believe, a major cost of our failure to confront the environmental crisis at its source, and to understand that only prevention works.

A major, if discouraging, example of how an understanding of its origin can lead to preventing the production of a pollutant is photochemical smog. For more than 20 years we have understood the origin of photochemical smog: The high-compression engines introduced after World War II to drive the enlarged American cars necessarily run hot; they therefore convert oxygen and nitrogen in the cylinder air to nitrogen oxides. Once out the exhaust, nitrogen oxides are activated by sunlight and react with airborne fuel and other hydrocarbons - many of them otherwise relatively benign - to produce ozone and the other noxious components of photochemical smog.

Can smogless engines that do not produce nitrogen oxides - that not only trigger smog but also contribute to acid rain - be built? They can; indeed, they have been. Every pre-World War II car was driven by such an engine; that is why the environment was then free of photochemical smog. In fact, nitrogen oxide production can be prevented without giving up the American car's precious over-powered engine (which is, nevertheless, a good idea). The so-called "stratified charge" engine can do just that. But manufacturers have thus far been unwilling to introduce them.

There are other examples of how pollution can be attacked at its source - and thereby prevented. Had American farmers been required to reduce the present, often unproductively high, rate of nitrogen fertilization - and to rely more on organic sources of nitrogen - nitrate water pollution would now be falling instead of increasing. If farms were required to shift from blindly repeated pesticide applications to integrated pest management, the rising level of pesticide pollution could be checked. If the railroads and mass transit were expanded; if the electric power system were decentralized and increasingly based on cogenerators and solar sources; if the pitifully small percentage of American homes that have been weatherized was increased - fuel consumption and the attendant air pollution could be sharply reduced. If bottlers were forbidden to use plastic bottles or to put plastic nooses on six-packs of beer; if supermarkets were not allowed to wrap polyvinyl chloride film around everything in sight and then stuff it into a plastic carrying bag; if McDonalds could rediscover the paper plate; if plastics were restricted to things in which they are really needed - let us say, artificial hearts or video tape - then we could push back the petrochemical industry's toxic invasion of the biosphere.

All this is strong evidence that environmental pollution originates in the technology of production and that the only effective way to improve it is to substitute a technology that eliminates the pollutant. This conclusion establishes a direct link to economic development. It identifies the factor - production technology - that not only determines environmental quality but is also the major determinant of economic development. What is at issue, therefore, is the relative impact of a given form of production technology on environmental quality on the one hand, and on economic development on the other. Are technologies that are more economically productive always more hazardous to the environment? Or, on the contrary, do economic and environmental merit go hand in hand, so that technologies can be found that are **both** economically productive and environmentally benign - and are therefore a means of solving the environment/development equation?

The conventional approach is that those technologies that are highly productive economically generally have a serious impact on the environment. This leads to the view that such technologies **must** be used as the means of economic development, so that environmental quality can then be achieved only by adding pollution control devices to them. In practice, developing countries are more likely to make a much greater effort to introduce economically productive technologies than to control their environmental impact - hence, the untoward result described in the earlier quotation from the Brundtland report. Unfortunately, as we have seen, this strategy has already failed in developed countries - the appended controls do little to improve the environment - and it offers no hope of solving the problem in developing ones.

Policy Strategies

The evidence summarized above dictates a new approach: that the only way to significantly improve the environmental impact of production to the **change** the forms of technology used to achieve it. The linkage between a technology's economic and environmental merit is therefore crucial. For example, if a new technology that is introduced in order to improve the environment also causes economic productivity to decline, once again there is a conflict between improving economic development and the environment.

One way to resolve this issue is to examine the linkage between the economic productivity and environmental impact of the new forms of production technology introduced after World War II. As noted earlier, in the period 1950-1970, there was a sweeping transformation in the technology of major aspects of production - that has disastrously degraded the environment. At the same time, these same technologies were responsible for a large increase in production and economic expansion. In operational terms, therefore, we can speak of the current environmental crisis as an untoward- and perhaps unexpected - consequence of the technological choices that were made in order to enhance economic development after World War II. It is this fact that often leads to the impression that economic development is necessarily accompanied by environmental pollution.

But this impression is misleading, for there is reason to question the **economic** merit of the post-war production technologies. Nuclear power provides a sobering example. It has been well established that economic development is closely linked to the availability of electric power, for it can be applied with nearly perfect efficiency to production processes ranging from tightening a bolt to driving a train. Clearly a bountiful supply of electric power can serve as the engine that drives economic development. Indeed, when nuclear power - the major post-war innovation in generating electricity - was introduced, it was hailed as an economic panacea. The head of the U.S. atomic energy program declared, for example, that "nuclear power will be so cheap that no one will bother to meter it". But the reality is very different. Nearly everywhere, the initial, rapid expansion of nuclear power has come to a halt. In the United States, for example, no new plants have been ordered in the last 12 years; many orders have been cancelled; nearly completed plants have been abandoned and even some completed ones are not allowed to operate; a power company heavily dependent on nuclear power has gone bankrupt, the first such failure since the Great Depression.

Nuclear power has failed as an instrument of economic development. The reason was clear even before Three Mile Island and Chernobyl. Nuclear power is an economic failure because of its **environmental** faults. The need to protect against accidents and routine radiation hazards has escalated the capital cost of nuclear power and turned it into an economic disaster.

Clearly, nuclear power is a particularly inappropriate vehicle of economic development in developing countries. Indeed, this is reflected in the experience of those developing countries that have mistakenly taken this course. Whereas among these countries the gross national product is proportional to the capacity of conventional electric power plants, this expected relationship is not exhibited by nuclear power capacity (Commoner, 1987b). This is direct evidence that nuclear power does not make an effective contribution to GNP - which is, after all, its purpose.

In nuclear power we have at least one dramatic example in which a technology's environmental defect gives rise to its economic failure, so that the two features are parallel rather than opposed. Nuclear power has failed so dramatically because the high costs generated by its environmental hazards have been **internalized** economically and are therefore directly reflected in its economic productivity.

There are more general examples that appear to be less dramatic only because, unlike nuclear power, the technology's environmental defects, although real, thus far remain economic **externalities** and do not yet appear in the industry's profit-and-loss columns.

The petrochemical industry is an instructive example. The enormous and increasing environmental hazards generated by this industry are only too well known. The petrochemical industry is equally famous for its economic success, having grown in the United States, for example, to a $250 billion industry in less than 40 years. What is less well known is that a serious effort to rectify the industry's environmental defects would literally destroy its economic viability. The petrochemical industry generates about

300 million tons of toxic waste annually, 99% of which is introduced into the environment in one way or another: deep-well injections, surface lagoons, storage tanks. Only one percent of the waste is destroyed, which is the only way to ensure that these often long-lasting and highly dangerous substances do not accumulate and eventually threaten living things. If the present (and still environmentally unsatisfactory) method of destruction-incineration - was applied to the active agents in the annual waste stream (about 1% of the total mass), the cost would be considerably more than the industry's total profit (Commoner, 1987a). In sum, the petrochemical industry is profitable only because it has managed, thus far, to avoid paying its environmental bill.

The environmental costs of other industries are more difficult to evaluate, but many of them may also be large enough to drastically unbalance the industries' books. How viable would the power industry be if it were required to pay the cost of acid rain - not to speak of the many other pollutants - that is produces? What would remain of the auto industry's already shaky profits if they were diminished by the cost of smog? And how would we reckon the net economic gain of modern industry after we confront the immeasurable cost of the flooding and climatic disruption that will ensue when the rising levels of carbon dioxide become critical - as they surely will if the numerous technologies based on fossil fuel consumption are not replaced?

Apart from these specific unmet costs, there are more general economic defects in the highly polluting post-war technologies. Most of them are large-scale, highly centralized capital- and energy-intensive enterprises. As a result, they are economically encumbered by low capital productivity (i.e., output per unit capital) and low energy productivity (i.e., output per unit energy). These economically burdensome factors are closely related to the harmful effects of these industries on the environment. For example, conventional power plants not only have a lower capital productivity than cogenerators, but, since they use more fuel per unit electric output, they are also a more serious source of pollution. In the United States, as in a number of other industrialized countries, it is depressingly clear that despite its past economic gains, the system of production is in a state of decline, exhibiting only very slight improvements in economic productivity in recent years. This economic problem may reflect, at least in part, the linked economic and environmental inadequacies of the post-war productive technologies.

Beyond these generic considerations, it can be argued as well that in and of itself public opposition to the environmental hazards inherent in these production technologies places a limit on their continued economic development. Thus, apart from the inhibiting effect of its high cost, the development of nuclear power - certainly in the United States and apparently in other countries such as Italy and, most recently, the USSR - is hindered simply because the public refuses to accept it at any price. Small but similar limitations on the continued economic growth of the petrochemical industry are beginning to develop in the United States: laws to ban the use of disposable plastic food packaging; demands for the banning of particularly hazardous chemicals; a growing competition from biological pest controls.

Thus, it can be argued not only that the post-war technologies are faulty environmentally, but that this very failing limits their continued ability to contribute to economic development. It appears, in sum, that the developed nations have been relying on production technologies that are severely limited in their ability to support further economic development **because** they also have harmful effects on the environment.

Based on these considerations, it is possible, I believe, to construct an approach to economic growth in developing countries that enhances development without imposing the environmental degradation that developed countries have endured. The basic precept can be stated quite simply, albeit negatively: developing countries should avoid the production technologies that have characterized the post-war production system in developed countries: centralized power systems, and nuclear power in particular; transportation based on high-compression internal combustion engines; agriculture based on the intensive use of synthetic chemicals; the petrochemical industry, almost in its entirety (excepting necessary and irreplaceable products such as medicinal drugs). Stated positively, the precept calls for the introduction in developing countries of those new technologies that correct **both** the environmental and economic defects-which, as we have seen, are closely linked - of the major productive technologies, which have caused so much trouble in the developed countries.

Energy production, which is such a crucial component of economic development, is a useful example of this approach. In those developing countries that have already introduced modern energy systems, they are largely based on the consumption of non-renewable fuels (chiefly, coal, oil, natural gas); they are also highly centralized, involving, for example, large capital-intensive facilities such as power plants and refineries. These features have generated both environmental and economic difficulties that can be avoided by adopting the opposite approach - that is, favoring renewable fuels and decentralized systems.

In practice this could be accomplished by a series of linked steps. To begin with, the need for electric power, at first necessarily based on non-renewable fuels, could be met by decentralized power plants based on cogeneration. Such plants recover both heat and electricity from the fuel; they are therefore more economic and less polluting than conventional power plants, which waste two-thirds of the fuel's energy in the form of rejected heat. For the sake of efficiency, cogenerators must be sized according to the local demand, avoiding the huge investments in a central power plant and its attendant large-scale transmission network.

Once such a decentralized, energetically and economically efficient energy system is in place, it can be gradually shifted from non-renewable to renewable energy sources. The cogenerator's conventional fuel can be replaced by solar fuels: ethanol produced from crops or vegetation, or methane produced from sewage and manure, or from marine algae. Finally, ethanol and methane can be gradually shifted to fuel motor vehicles, and photovoltaic cells used to produce electricity, augmented by solar collectors for heat. In each case, these technologies sharply reduce environmental impact, in comparison with the conventional ones, cut the cost of energy derived from non-renewable fuels, and eventually free the economy of the self-destructive effect of the ever-increasing cost of non-renewable fuels.

The importance of a new departure in agricultural technology in developing countries, and its relation to the petrochemical industry, is emphasized by the recent disaster at the plant in Bhopal, India, that manufactured insecticides. The conventional remedy suggested by this catastrophic event is that since insecticides are essential to Indian agriculture, the plant must continue to operate, but that a greater effort must be made to prevent such accidents and to protect the surrounding communities from them if they occur. More to the point, the disaster raises a question about the need for the insecticide, and therefore for this inherently dangerous plant. It has already been demonstrated in the United States, for example, that the substitution of biological controls and integrated pest management for the conventional heavy applications of insecticides greatly reduce costs and environmental impact. Net returns to farmers have been increased by more than $500 million (Brown et al., 1988). In fact, integrated pest management and biological controls have already been applied to certain crops in developing countries. Excellent pest control as well as economic improvement have been achieved (Brown et al., 1988).

There is no doubt that the Brundtland report has been a very valuable reminder of the urgency of the environment/economic development problem. Quite properly, it has emphasized the complexity of the problem and the conflicting interests that hinder any proposed solutions. I believe that the time has come to take the next step and define what can now be done to end the tragic impoverishment of so large a part of the world without imposing on it an unwelcome price - a degraded environment. We now know, I believe, how that next step can be taken. We need to choose, as instruments of production, those technologies that are economically productive because they are environmentally benign.

This international social goal requires social governance of the means of production. Yet, the exclusive power to govern productive decisions is jealously guarded by firmly entrenched ideologies that are often regarded as immune to debate. The lesson of the environmental experience - an experience of failure thus far - is that the time has come to expose these ideologies to the enlightenment of open debate.

References

Brown, L.R. et al. (1988), State of World, 1988, Worldwatch Institute, New York: W.W. Norton & Co.

Brundtland, Gro Harlem, World Commission on Environment & Development (1987), Our Common Future, Oxford: Oxford University Press.

Commoner, B. (1972a), The meaning of the environmental crisis, Miljoforum/Environmental Forum, Presented at the Environmental Forum, U.N. Conference on the Human Environment, Stockholm, June 5.

Commoner, B. (1972b), The environmental cost of economic growth, Chemistry in Britain 8:52-65, Feb. See also: The Closing Circle, 1971, New York, Alfred A. Knopf, Inc.

Commoner, B. (1987a), A Reporter at Large: The Environment, New Yorker, June 15. See also: Il cerchio da chiudere, 2nd ed., 1986 (Milan: Garzanti Editore).

Commoner, B. (1987b), Nuclear power for the Third World: Bane or blessing: World Policy Journal, Spring.

Smith, R.A. et al. (1987), Water-quality trends in the nation's rivers, Science, vol. 235, 1607-1615, Mar. 27.

U.N. Economic Commission for Europe (1987), Environment Statistics in Europe and North America, Conference of European Statisticians Statistical Standards and Studies, no. 39., New York: United Nations.

U.S. Dept. of Commerce (1985), Business Statistics 1984, 24th Edition, Bureau of Economic Analysis, Washington, D.C.

U.S. Dept. of Commerce (1986), Statistical Abstract of the United States, 1987, 107th Edition, Bureau of the Census, Washington, D.C.

U.S. EPA (1984a), Environmental Quality, The 15th Annual Report of the Council on Environmental Quality, Washington, D.C.

U.S. EPA (1984b), Lake Erie Intensive Study 1978-1979. Final Report, prepared by D.E. Rathke, Ohio State Univ., Columbus, OH, EPA-905/4-84-001, Jan.

U.S. EPA (1986), National Air Quality and Emissions Trends Report 1984, Office of Air Quality Planning and Standards, Research Triangle Park, NC.

U.S. EPA (1987), National Air Quality and Emissions Trends Report 1985, Office of Air Quality Planning and Standards, Research Triangle Park, NC.

Chapter 10

COMPREHENSIVE SOCIAL ASSESSMENT: AN ESSENTIAL INSTRUMENT FOR ENVIRONMENTAL POLICY-MAKING

F. Archibugi

Prologue

The use of Environmental Impact Assessment (EIA) has become increasingly widespread during the last few decades, because environmental conditions have been identified as critical factors in determining our well-being and quality of life. The objective of this chapter is to underline that EIA is reliable and useful only if developed within a comprehensive approach to planning and evaluation.

The conclusion of our analysis recommends the incorporation of EIA in a **more general process of evaluation and decision**, which guarantees that environmental policy decisions are evaluated within the complexity and comprehensiveness of development planning.

In the present chapter a number of general considerations on the importance of comprehensive planning procedures will be developed in order to underline the indispensability of these procedures in turning environmental policy from a "negative" strategy into a "positive" one.

Comprehensive Assessment and Environmental Impact Assessment

The EIAs which have been developed until now are usually linked to single projects. Their reliability is often questionable, which is mainly due to the paucity of national parameters.

(1) Analogy with cost-benefit analysis experience

The limited reliability of EIAs runs parallel to Cost-Benefit Analyses (CBA) of projects. Even though such analyses had a different scope (which in most cases excluded the environmental costs or the environmental effects, since they are difficult to quantify according to the units of measurement used in CBA), they had -and continue to have- the defect of not being "hooked up" to an adequate system of national parameters such as the relative weight of economic objectives related to the national economy, the social discount rate, the shadow price of investment, the shadow wage and the shadow price of foreign exchange. In fact non-existence of central planning systems which might be able to provide these parameters (and make the evaluation of costs and benefits of single projects coherent and credible) was certainly a serious handicap to the development of suitable CBAs.[1]

In fact, the need for these parameters presupposes the existence of a policy agency competent to formulate them and to use its appropriate decision-making power to bring them to a decision.[2] The lack of an adequate process of comprehensive planning - and of its capable management - thus implies the lack of adequate national parameters as well. The lack of national parameters, in turn, implies that any project analysis ought to construct and presuppose, case by case, its own reference parameters. Assessments which are done quickly exhibit various precarious conditions, and their limitation is often apparent in the whole exercise.[3].

And finally, such a method of doing a CBA without a reference situation in order to put it into context, runs the risk of being strongly manipulated due to the single-faceted and often partial view of the project planners and the associated analysts in each project.[4]

(2) The indispensability of national parameters for EIAs

Environmental Impact Assessment shows a similar picture. Here too, approaches are often on the level of a single project, whilst adequate national parameters are lacking. The question is: What are the parameters that are appropriate for an "environmental" cost and benefit analysis in which the value and the criteria **are not quantifiable** in traditional economic terms (i.e., market prices or shadow prices)?5

The national parameters in this case are those of "value", as attributed in any manner (either qualitatively or quantitatively) to the physical/natural environment (including the urban environment). The parameters that are related to valorization and devalorization from the viewpoint of the national community are eventually articulated according to the viewpoints of its diverse groups and interests·6

Obviously this assessment can only be a political one, eventually assisted by a technical predisposition.[7]

In order to be expressed in an adequate manner, and not in an occasional, casual or chaotic one, this political assessment ought to follow precise decisional procedures which will make it ordered and effective.

Land-use Planning and Indicators of Environmental Quality

In order to achieve the determination of indicator values assigned to the physical/natural or regional environment, there are only two possibilities:

(i) First, an identification of the entire area of interest to the reference community (whether regional, national or global) of the **objectives of appropriate use** for the area, taking into consideration its diverse geographic areas, taxonomically distinct one from the other. We shall call this identification the construction of a **Land-Use Planning Framework** (LPF).[8]

(ii) Afterwards, an identification of the relevant phenomena and/or factors of environmental quality in each of the "niches" that the LPF has identified for an appropriate use of the territory itself (fixing also the threshold values for these factors and using a unit of measurement appropriate for each one). We may call this identification an **Environmental Quality Indicator System** (EQUIS) related to the environments or areas of the LPF; and we shall call the thresholds or the pre-selected values **norms** or **standards** of those factors of environmental quality.[9]

An example of the logical sequence for the undertaking of the two aforementioned modalities of evaluation may serve as an illustration: the task of the LPF (for the territory in question) will be to identify, given the present land-use, **what areas will be designated** (because of their best quality for that use) as urban centres. For these centres the LPF will indicate the concrete delimitation in order to distinguish the typological areas for a policy of environmental conservation as well as to determine the characteristics of density, functionality and role. Having determined this, the tasks of EQUIS would be to fix the indicators which can express the environmental quality threshold for such areas.

Let us suppose that among these indicators the EQUIS indicates the amount of parking space available for registered vehicles. The norm of the environmental quality will therefore be the increase of x m^2 considered as an indispensable minimum threshold needed to guarantee an acceptable (urban) environmental quality.

Plan Evaluation

Environmental quality - in public opinion - is highly influenced by appropriate land-use. For ages appropriate land use, whether urban or rural, has been the objective of land-use planning. Every "blueprint" plan after all aims to assign an appropriate use to each area. From this point of view, it can be stated that a good conservation and assessment policy is the same as a good land-use plan, and vice versa.[10]

But "blueprint" planning only recently has contemplated the possibility of introducing systematic methods, and not only intuitive ones, of assessing alternative land-use choices. Until now blueprint plans did not include alternatives, nor systematic assessments. These assessments were the result of the planner's intuition which, at best, used some method of calculation as a support.

More recently the idea that an area can have multiple uses has been introduced. In order to choose between the uses it is possible to adopt several rational and systematic methods, so that errors of assessment and arbitrariness can be avoided. This is known as "Plan Evaluation".[11]

Plan Evaluation, precisely because it is less unilateral is a "noble relative" of EIA: it is "noble" because it includes a "comprehensive" assessment of all the possible choices at play.[12] EIA remains extremely important as a source of information of a technical nature (as "analysis" rather than "assessment"). Other sources of information on environmental, social, economic, functional and institutional impacts are equally important and should be present in a "comprehensive" integrated assessment.[13]

Plan Evaluation is assigned to comprehensively evaluate **all** the possible costs and **all** the possible benefits, quantifiable as well as non-quantifiable. It is a way of proceeding to an effective decision which at the same time is the most rational and realistic one.[14]

Single-faceted approaches (and the methods which support them, such as CBA and EIA) do not produce effective decisions by themselves. In the case of CBA, it does not because it emphasizes only quantifiable costs and benefits, and not unquantifiable ones; sometimes even only the "internal" costs and benefits and not all "external" ones (as any proper environmental philosophy rightly denounces). In the case of EIA, it tends to highlight only the environmental costs (or benefits) and not costs and benefits of any other type. Their single-facetedness is often, understandably, the very reason that a decision cannot be made; and assessment begins to be feared as an obstacle, rather than seen as a support to the decision-making process.

Single-faceted approaches are obviously less rational (often "sub-optimal") as they do not take into account all the factors which intervene in the action and/or in an assessment. They seem at first view more realistic because it is in single-faceted terms that the planned developments are studied and proposed. But this apparent realism is illusory because it is completely supplanted by a "stalemate" produced by conflicting interests. These cases need quite a bit more energy and time to settle things and get out of the impasse they created, than those that began with a multifaceted, integrated approach. The latter incorporate diverse points of view within the decision-making process.[15]

A Method of Allocating Values Related to Land: The Land-Use Matrix

The comprehensive assessment of various alternative land uses (or Plan Evaluation) implies the capacity to assign a **use value** to land (which is commonly defined as a non-renewable public asset) without forgetting a certain shadow-value of exchange.

For this we proposed above the construction and the utilization of a Land-Use Matrix (LUM), in order to assign a value to each portion of the land in use, or under consideration for use, in relation to the **use demand** and the **supply of available land.**[16]

The LUM consists of an "ideal type" Table in which:

(i) in the rows (or columns) the available quantity of land for every fixed category of land use is indicated, the taxonomy of which is based on its quality (seen from the point of view of its possible uses)

(ii) in the columns (or rows) the quantity of use demand based on present activities (production consumption activities) coming from a given population or the **future** activities as forecast by programmes or projects or interventions elaborated upon in the plan) is indicated.

Land, as a non-renewable public asset, can acquire different values for the community based upon the amount of land available, the amount of appropriate use demand, and on particular circumstances related to the level of satisfaction of diverse needs which determine the preference model of the community in question with respect to a comprehensive needs schedule.

When we say "appropriate", we are introducing a concept of the quality of the land-use in relation to the area's characteristics, a concept which can be linked to that of assigning values.

In the framework of land-use planning, it is taken for granted that the future land-use demand (which in the LUM is compared to the available supply of land) will be appropriate.

In the creation of a matrix of present land use and supply, the quantity of appropriate and inappropriate use can be seen in each cell of the matrix and thus a conventional value can be assigned to each of them. If the plan also has the objective of restoring part of a region which has been used improperly, to an appropriate use (as, for example, in some urban renewal and regional renewal projects), the difference between the values constitutes an estimate of the operation's net gains.

The quality of the land resource which should be recorded in the matrix can also come from other characteristics or properties which are capable of determining or altering the value, for example, rarity in itself (and not simply as it results from the relationships between supply and demand which the matrix shows), or the degree of non-renewability, and so on.

The values which are generated constitute the **national parameters**, on which calculations of gain and loss from alternative land use projections can be done, either in regional plans, programmes or single projects.

Towards Integrated Environmental Planning: Some General Considerations

At this point we will review the discussion on the opportunities and even the necessity of an integrated approach to assessment (and planning). We shall reinterpret the relationship between development assessment and environmental protection assessment.

(1) The economic-oriented approach

Once there was a science of economy (economics) directed to unveiling the secret mechanisms of development (such as the quantity of goods and services available); and to define and establish (for the government) prescriptions and norms capable of making development as elevated and as fast as possible.

When economic science began to be used to judge the opportunity - in order to obtain the most elevated and quick development possible - of operations, programmes and projects (public or private) in themselves or as alternatives, economists began to look for ways of elaborating methods of analysis and evaluation of the contribution of each one to development. Development was considered as an increase in utility or productivity. The latter qualities, in turn, were considered to be the relationships between employed resources and the results obtained, or between tolerated prices and benefits gained. The greater the net difference produced between these two variables (and the faster they became evident), the greater the advantage seen in the adoption of an operation, a programme of action or a project. CBA has thus become the rational premise for public operations, whether direct or indirect, for the formation of new products (material goods or services).[17]

As stated earlier, CBA for development projects is considered too single-faceted, and overly oriented towards the single criterium of **economic** development, understood to be the balance between input and output in the productive process. CBA is not sufficiently flexible to consider other criteria, non-economic objectives or other values: for example, values which are difficult to quantify, such as environmental conservation. If the latter criteria, objectives and values are somehow taken into account, they would seriously modify the results of such analyses and assessments. CBA, in so far as it is a rather single-faceted analysis, has not developed the capacity to effectively orient the behaviour of politicians or of the operators.

(2) The ecology-oriented approach

Later on, ecologic science tried to unveil the secret mechanisms of natural and environmental relations. It was thought that one would apply the physical rule that in nature "nothing can be created or destroyed", to economic relations. It was concluded that every added value was equalized, in some fashion, by a subtracted value and that the appreciation of the value added (the net production) could not be disassociated from the consideration and computation of the value subtracted, as with natural resources, which once used up, are non-renewable or, more generally, any negative revenue which cannot be recorded on the goods and services market.

Subsequently attempts were made to assess planning and projects according to the impacts they had on nature and on non-renewable resources, and to accept only those operations which did <u>not</u> have a negative environmental impact.

Later it was realized (perhaps too slowly) that even this type of assessment is excessively single-faceted, is too narrowly focused on the absolute defence of nature **telle quelle**, has no reference to its social utility, to its use for human or social interests, and sacrifices other criteria, objectives and values.

Well known economists (for example Clark, Mishan and others) have criticized the logic of judging any operation in terms of growth maximization, complaining about a "growth-mania";[18] the day well-known ecologists do the same, and coin "ecology-mania" will be a step forward for the critical spirit, rationality and finally, for good sense.

(3) Overcoming partial approaches

Unfortunately the two above mentioned approaches have difficulty finding an adequate point of encounter and integration. They seem rather to accompany (whether as cause or as effect is difficult to say) a radicalization of the fronts; they seem to have developed as opposite approaches, conflicting rather than cooperative.

And yet for some time it has been recommended that an approach be used to integrate CBA and EIA into a "superior" specimen of comprehensive assessment. In this type of assessment all criteria, objectives and values would be accounted for, without necessarily favouring one to the detriment of another, but rather assuring correct trade-offs. Evidently, as for all great reforms of thought or action, we must wait patiently until the cultural level of technicians and politicians matures with time.[19]

(4) The planologic approach

This latter method, from a politico-economic point of view, is referred to as the **planning process**. Since it deals at the same time with all the substantial aspects of human social welfare, it is an integrated socio-economic planning process (comprehensive, unified, global or systematic). More specifically if it is concerned with the relationship between economic and environmental well-being, then it is the comprehensive land-use planning itself.

The same method, from a technical standpoint, is called **multiple criteria (or objective or value) assessment methods and techniques**. These are the methods and techniques which support decisions and planning.[20] As with many cognitive instruments which have been widely proposed at the scientific level, they prove difficult to apply to the adequate political process of planning at the operational level.

The recognition and acceptance of both methods have proven difficult. As mentioned previously the largest impediment is the cultural level of the operators (politicians and technicians) who all tend to refute what they do not know, and that which does not rapidly reach a consensus. Perhaps another factor of resistance is the wish of some interested and privileged powers not to introduce rational methods of decision-making which, by their very nature, tend to clash with the arbitrary exercise of power. However, it is also true that in the most advanced and cultivated circles of this same power, the planning methods and decision-making techniques we are referring to already find great acceptance. Thus we maintain that the most convincing explanatory factor of this situation is still the cultural factor.

From the NO Strategy to the YES Strategy by the Process of Integrated Planning

In fact, we wish to combine the varied and manifold aspects which must confront, on the one hand, a political economy which wishes to be geared toward a defence and assessment of the environment, and, on the other hand, an environmental policy which does not wish to hinder the realization of the goals of economic and social well-being. In this context, it is necessary to introduce methods of integrated planning, viz., by adopting the planologic approach.

It is difficult to imagine that the environmental policy approach which until now has mainly been negative, can be overturned in other ways.[21] To do so the world of politics must demonstrate a great organizational and reformational capacity - which is the truly lacking technology of the present day. In fact, the awareness of our problems is not a shortcoming in the world of modern informatics. What we do not know well is merely how to manage them.

Technological potential comes from science and technological knowledge. But there is quite enough of that; some go as far as saying there is more than enough. What we are lacking, rather clearly, is a **technology of political and social action**, which could be made to govern science and technology with tranquility and safety. And also, where this technology of political action is already available - as in the case mentioned earlier -, it is difficult to find a way to apply it.

Our organizational and controlling capacity with respect to the entropy of development, both scientific/technological and political, remains mainly "negative".

In this vacuum of political decision the entropy of natural forces is at its greatest. The freedom of nature ungoverned by man is at its greatest. But, also at its greatest is the risk of catastrophe.

Mankind cannot leave the duty of resolving these problems and guaranteeing security merely to ecological, naturalistic, evolutionistic or biological "equilibrium", no matter how surprising and wonderful they may be to the scientist or the nature lover.

To do this, man would be negating himself. His progress has always been founded upon social and moral progress. His consciousness has always been founded on his liberty. To make his freedom of collective decision and of rational choices subject to naturalistic determinism, and scientific positivism, is exactly the positivist illusion which for the last two centuries accompanied both the most grandiose scientific and technical advances and the most uncontrollable political degradation, holocausts and social and environmental risks. Unlike in nature, there are no immanent laws for social life to subject society to a "managed" evolution.

At a more advanced level, we must recuperate the anxiety and the reform-oriented and constitutive willingness of political and social

Enlightenment, which furnished the last substantial political scheme upon which modern society has founded its general historical progress - that of democracy, of participation and of social and economic liberty. And constitutive reform which today is spoken of everywhere, at the national level and above all at the global level, cannot be large enough. Minor issues which flaunt being pragmatic and realistic are a hindrance to the greater problems which political society must face in order to get in step with the great scientific, technical and economic progress of our century and of the social impact that resulted from them. To be useful these reforms must become finalized at a global level as well as at the national level in order to introduce new useful **planning systems integrated into the decision-making process**. Clearly, this is the most important challenge to which contemporary political structures must respond.

Notes

[1] "The efficacy of social CBAs depends on how one determines and uses national parameters" (Dasgupta, Sen and Marglin, 1972, para 1.3). "The success of project formulators and evaluators obviously depends on the degree to which the first approximations of the parameters accurately reflect the national priorities. If these are "prices fixed by chance," to use one of Walras' terms, project formulation and evaluation quickly loses its significance... And it is at this point that ... one understands why the link between project formulation and evaluation and national planning is essential." (Dasgupta et al., Ibid., par. 11.4).

It has been accepted for some time that the social profitability of projects must be ascertained on the basis of an identification and enumeration of costs and benefits from a social point of view rather than a private one, and that such profitability must be based on an evaluation according to shadow-prices which reflect assessments or social preferences rather than private preferences.

But how important it is to distinguish between "**Pareto social optimum**" which reflects the point of view that one must take into account individual preferences and the "complete" function of social well-being based on the notion that society is "an organic chooser of ends," to use an expression of Amartya Sen (1970) is rather less accepted. Sen defines **Pareto** ranking as a "quasi ranking" and organic social ordering as "ranking". Papandreou (1970) defines the first as "partial ranking" and the second as "complete ranking."

The CBA which has predominated in politics for the past twenty or thirty years is one which developed without the choices or pre-ordered plans and which is related to the first notion (**Pareto "incomplete"**) of social well-being. But the lack of a coupling to a "complete" idea of social well-being - that of which is related to the appropriate definition of national parameters - results in not only an incomplete assessment of projects and plans, but also in an assessment which is highly unreliable (cf., also Weisskopf and Marglin, 1969).

[2] The essential role of a central planning agency with regard to the CBA is largely described in the work cited by Dasgupta, Sen and Marglin (1972) which, as is a UNIDO manual, instead of the title "Guide to Project Evaluation" would merit rather the title "Guide to the Formulation of Criteria and Methods for National Planning Capable of Making Single Projects Evaluation Possible and Reasonable."

The implications for project formulation and the assessment of situations in which a social, "complete ranking" is spelled out for alternative economic states (that is, in situations where all decisions are made in the absence of a society which is an "organic chooser of ends") is in an important (and little known) essay by Papandreou & Zohar (1971).

In this essay in particular there is an interesting development of the relationship between "national plan" and "programme" (defined as a set of projects whose impact on the economy is indecomposable, and single "projects (defined as a "sequence of activities" which merits the appellation of projects if and only if it is "indecomposable" (that is to say, if and only if it cannot be expressed as the sum of two "subseqences").

[3] "If the assessors are not required to apply fixed social parameters and values, their judgement can constitute an important factor towards a choice which could cause a state of chaos..."

"It is worth remembering that the CBA is not a technique, but an approach. It furnishes a rational context to the choice of projects by using national objectives and values. Projects are judged in terms of their precise effect on the national economy, and this effect is measured using parameters which reflect national and social objectives" (Dasgupta et al., 1972, par. 1.4).

[4] "A project assessor can practically justify any project starting from adequate pre-suppositions such as: tax on interest, consumption habits, gains values which are obtained in foreign currency, etc." (Dasgupta et al., Ibid).

[5] On the limits of the CBA as applied to environmental projects, see Baumol and Oates (1971) and Pearce (1976).

[6] But ultimately considered as "an organic chooser of ends".

[7] As appropriately affirmed in Dasgupta et al. (Op. cit. par. 11.3), referring to the CBA, but which also seems valid in cases of environmental projects when the qualitative judgement is by now more arbitrary than the quantitative one. "National parameters which represent the relative weight of diverse objectives, and shadow-prices, which are instruments that contribute to the achievement of these objectives, must reflect conscious political decisions about issues that are, after all, political issues. To

allow to project formulators and assessors to determine national parameters is to leave political decisions to technicians and deprive politicians of the role they are expected to take in the decision-making process."

[8] Past demonstrations of elaborations of the LPF at many levels of political signification and authority are considered in: "Proiezioni territoriali" of Project 80 (Italian Ministry of the Budget and Economic Planning, 1969-71); "Schema d'Amenagement de la France" of DATAR (DATAR, 1968 and thereafter); and "Raumordnung Programme" of the Government of the Federal Republic of Germany (FRG, Federal Ministry of Regional Planning, Building and Urban Development, Regional Planning Programme for the Spatial Development of the Territory of the FGR, 1975).

The negative aspect of such scenarios (except the German one which seems to have significantly influenced the operation of federal investments) was that of being concluded with their publication and not to have had any permanent follow-up of application, revision, updating or improvement.

[9] A system of physical quality indicators of the receptors of environmental pollution (water, air, ground) is more or less present in every country where ad hoc institutions for the monitoring and control of environmental quality and the relative official periodic reporting of these indicators exist. However, indicators of the value (quality) of alternative uses of regional resources are absent; such indicators are of the type which spring forth from a matrix of availability and use of land (which will be discussed in section 4) Furthermore, indicators are absent also, more or less "synthetic, about the quality of the urban environment as such, which has crucial importance for an environmental policy directed to a human and social settled well-being.

[10] For a systematic treatment of the relationship between land-use and environmental quality, it is worth looking at the classic treatise of Chapin (1965), in particular in the 3rd Edition with Kaiser (1979) and especially Chapter 7 ("The Users of the Land: Their Activity Systems and Choice of Space Qualities," pp. 194-230); Chapter 8 ("The Use of Land: Development and Developing Areas," pp. 231-288); and Chapter 9 ("Natural Environmental Inventory and Analysis, " pp. 289-326). Another systematic treatment by Archibugi (1982) should also be mentioned, especially Chapter 5 ("Analysis and Assessment of Land Resources"), Chapter 11 (The Land-Use Policy"); Chapter 21 (The Protection of the Natural Environment"); and Chapter 29

("The Prospective Balance of Land-Use").

[11] Already the passage from the assessments of "projects" to that of "plans" by itself constitutes an important step towards multi-faceted integration, treating the "plans" generally many projects and in such a way many objectives relative to a given territory taken as a whole. If the "multiple criteria" approach is already an important advancement for "assessors", which are usually applied to single projects, it is less so for planners, who are already used to having a "comprehensive" approach, i.e., an approach oriented towards many aspects and many projects, for a given territory. As the regional maps scale increases, the problems of environmental valorization and protection become more and more deeply ingrained in the project or planning process (and becomes therefore of less importance for the "environmental impact" to be taken into account, since environmental valorization and protection are already among the principal objectives of every plan).

[12] Therefore, in the first place among these is environmental valorization.

[13] For a panorama of the many directions of "social" valuation it is worth considering an anthology edited by Pearce (1978).

[14] Plan evaluation began attempting to adapt CBA to environmental problems. Diverse methods were proposed:

- the "shadow-project approach" of Klassen (1973) and Klassen and Botterweg (1973 and 1976); the shadow-project is carried forward simultaneously with the basic project in order to compensate in real term the damage to the environment from the basic (socio-economic) project;
- "cost-effectiveness" analysis: see among others English (1968) and Seiler (1969), in which costs are directly correlated to levels of fixed objectives;
- "threshold analysis" (see Kozlowski, 1968), Malisz, 1970 and Kozlowski and Hughes, 1972), founded on the fixed supply to which are connected continuous curves of demand, among which evaluators identify the single "optimal" one (that is the threshold);
- plan evaluation was finally faced with (at the confines of the use of monetary and non-monetary instruments) the proposal of the "planning balance sheet method" of Litchfield and others (1968 and 1975); this

method includes - where possible - effects expressed in monetary terms on a certain number of various "social" sectors; and if it is not possible in monetary terms, includes effects expressed by other units of measure or other indicators of an ordinal or nominal scale;

- another approach to evaluation is that of the "participation method" founded upon permanent discussions between or among interested parties, in which at the moment of selection are emphasized and brought up for discussion (see for example, Manheim et al., 1974 and 1975);
- the multi-attribute utility theory is not to be undervalued as it has offered methods to evaluate a number of possibilities of explicitly formulated choices (Fishburn, 1970 and Keeney and Raiffa, 1976);
- plan evaluation finally gave form to diverse methods of multi-criteria evaluations, which recently are multiplying and merit a systematic illustration in order to be correlated taxonomically to the diverse problematics and the diverse decisional processes. Among the first approaches to multi-criteria planning must be mentioned the noted one of Hill (1973). A good survey of these methods is in Voogd (1983).
  A panorama of the "State-of-the-Art" is in the collection of writings in Fandel & Spronk (1985).

[15] The multi-criteria approach to decision-making, which is rich in a larger variety of "schools" and relative methods and "techniques" originated within the field of "Operational Research". It certainly broadened its own area of interest with respect to the CBA, when the area of transactions (and related decisions) based on "non-quantifiable values" is enlarged; thus rendering more difficult evaluations through the attribution of figurative shadow-prices of an "availability to paying". The entire problematic is trated with an abundance of examples in Sinden and Worrel (1979).

[16] On the LUM see Chapters 5 and 29 of the work cited: "Principles of Regional Planning" (Archibugi, 1978) as well as the recent contribution to a Seminar at the University of Rome (Archibugi, 1988).

[17] For an up-to-date excursus from the first Pigouvian schemes on the welfare economics (1920) to the most modern theories on the measurement of "social costs" see Nash's panoramic text (1978).

[18] See Clark (1961), Scitovsky (1964), Mishan (1967) and Daly (1973).

19 From the vast "planologic" literature we shall cite a few particularly significant and basic works. First of all several fundamental papers of Ragnar Frisch, written (but never as widely discussed as they ought to have been) and published posthumously (Frisch, 1973); among these, one, on "Cooperation between Politicians and Econometricians on the Formalization of Political Preference" (1970) and another on "Implementation System for Optimal National Economic Planning without Detailed Quantity Fixation from a Central Authority" (1964) presented in Rome during the First World Congress of the Econometric Society. Then a short piece by Wassili Leontief (1977) on what he means by "economic planning", and on the informative instruments and basic procedures upon which it can function at the national level in a pluralistic society with public power and private liberty. Lastly, the two volumes of Leif Johansen in which his "Lectures on Macro-Economic Planning" (Johansen, 1977-79) are collected, in which nearly all the basic methodologies for the construction of a planning system are presented. Other useful and systematic contributions are those of Caire (1972) and several papers of Archibugi; (collected in Archibugi, 1979). An important technical operative on the problems of general planning was anticipated in a well-known conference hosted by OECD in Bellagio (Lago di Como) in 1968, (for which, see the numerous contributions of Ozbekhan, Forrester, Jantsch, Rea, etc. in OECD, 1969). Further bibliographic references on the evolution of planology can be found in the bulletin of "Sistema informativo sulla scienza della pianificazione," put into operation by the **Centro di studi e piani economici di Roma** (Rome "Planning Center") with the support the **Consiglio Nazionale delle Richerche** (National Research Council) beginning in 1985.

[20] An important survey of these methods is in Nijkamp (1977, chs. 8 and following).

[21] To drive it towards what Giorgio Ruffolo several years ago called "creative ecology" (Ruffolo, 1985, pp. 165-167).

References

Archibugi, F. (1979), Verso un Sistema centrale di Pianificazione. Contributi vari per una Metodologia generale della Pianificazione, Centro di studi e piani economici, Roma.

Archibugi, F. Principi di Pianificazione Regionale, Angeli, Milano, 1979 (1 ed.), 1982 (2 ed.).

Archibugi, F. La Matrice di Uso del Territorio, strumento di analisi e di valutazione ambientale, Relazione al Seminario del Dipartimento di Pianificazione Territoriale e Urbanistica, Università di Roma, La Sapienza, 18-3-1988.

Baumol, W.J., and Oates, W.E. (1971), The Use of Standards and Prices for the Protection of the Environment, Swedish Journal of Economics, vol. 73.

Caire, G. (1972), La Planification; Techniques et Problemes, 2 ed., Cujas, Paris.

Chapin, F.S. (1979), Urban Land Use Planning, University of Illinois Press, Urbana, 1 ed. 1965, 3 ed. (with E.J. Kaiser).

Clark, C. (1961), "Growthmanship". A Study in the Mythology of Investment, Institute of Economic Affairs, London.

Daly, H.E. (1973), The Steady-State Economy: towards a Political Economy of Biophisical Equilibrium and Moral Growth, in: Daly, H.E. (ed.), Toward a Steady-State Economy, Freeman, San Francisco.

Dasgupta, P., Marglin, S. and Sen, A. (1972), Guidelines for Project Evaluation, UNIDO, Wien.

Delegation Generale a l'Amenagement du Territoire et a l'Action Regionale (DATAR), Schema d'Amenagement de la France, (plusieurs volumes), Paris, 1968 et suivants.

English, M.J., (ed.) (1968), Cost Effectiveness, Wiley, New York.

Fandel, G. and Spronk, J. (eds.) (1985), Multiple Criteria Decision Methods and Applications, Springer, Berlin.

Federal Republic of Germany, Federal Ministry of Regional Planning, Building and Urban Development (1975), Regional Planning Programme for the Spatial Development of the Territory of the FRG ("Raumordnung Programme"), Bonn.

Fishburn, P.C. (1970), Utility Theory for Decision Making, Wiley, New York.

Frisch, R. (1976), Economic Planning Studies, (Selected and Introduced by F. Long, Preface by J. Tinbergen), Reidel, Dordrecht-Holland.

Hill, M. (1973), Planning for Multiple Objectives, Regional Science Research Institute, Philadelphia.

Johansen, L. (1977-1978), Lectures on Macro-Economic Planning, 2 vols. North-Holland.

Keeney, R.L. and H. Raiffa (1976), Decisions with Multiple Objectives; Preferences and Value Trade-offs, Wiley, New York.

Klassen, L.H. (1973), Economic and Social Projects with Environmental Repercussions: a Shadow Project Approach, Regional and Urban Economics, vol. 3 no. 1.

Klassen, L.H. and Botterweg, T.H. (1973), Evaluating a Socio-economic and Environmental Project, Netherland Economic Institute, Series: Foundations of Empirical Economic Research, vol. 9.

Klassen, L.H. and Botterweg, T.H. (1976), Project Evaluation and Intangible Effects; a Shadow-Project Approach, in: Nijkamp, P. (ed.), Theories of Environmental Economics, Nijhoff/Kroese, the Netherlands.

Kozlowski, J. (1968), Threshold Theory and the Sub-regional Plan, Town Planning Review, July.

Kozlowski, J. and Hughes, J.T. (1972), Threshold Analysis: A Quantitative Planning Method, Architectural Press, London.

Leontief, W. (1977), The Economic Planning: Methods and Problems, in: Essays in Economics. Theories, Facts and Policies, Sharpe, White Plains.

Lichfield, N., Kettle, P. and Withbread, M. (1975), Evaluation in the Planning Process, Pergamon, Oxford.

Malisz, B. (1970), Direct Application of Threshold Analysis to Regional Planning, Paper from the Seminar on Threshold Analysis, Centre for Environmental Studies, London.

Manheim, M.L. (1974), Reaching Decisions about Technological Projects with Social Consequences: a Normative Model, in: Systems Planning and Design, De Neuville and Marks, Prentice-Hall, Englewood Hill.

Ministero del Bilancio e della Programmazione Economica (1969-1971), Le Proiezioni Territoriali del Progetto 80, 3 vols., Roma.

Mishan, E.J. (1967), The Cost of Economic Growth, Staples, London.

Nash, C. (1978), The Theory of Social Cost Measurement, in: The Valuation of Social Cost, D.W. Pierce, Allen & Unwin, London.

Nijkamp, P. (1977), Theory and Application of Environmental Economics, North-Holland, Amsterdam, 1977.

OECD (1969), Perspectives of Planning, Paris.

Papandreou, A.G. and Zohar, U. (1971), Programme-Project Formulation: Evaluation and Selection in the Context of a National Plan, Economics of Planning, vol. 11, no. 1-2.

Pearce, D.W. (1976), The Limits of Cost-Benefit Analysis as a Guide to Environmental Policy, Kyklos, 1.

Pearce, D.W.(ed.) (1978), The Valuation of Social Cost, Allen & Unwin, London.

Ruffolo, G. (1985), La Qualita Sociale. Le Vie dello Sviluppo, Laterza, Bari.

Scitovsky, T. (1964), What Price Economic Growth, in: Papers on Welfare and Growth, Stanford University Press, Stanford.

Seiler, K. (1968), Introduction to Systems Cost-Effectiveness, Wiley, New York.

Sen, A.K. (1970), Collective Choice and Social Welfare, Holden-Day.

Sinden, J.A. and Worrel, A.C. (1979), Unpriced Values: Decisions without Market Prices, Wiley, New York.

Voogd, H. (1983), Multicriteria Evaluation for Urban and Regional Planning, Pion, London.

Weisskopf, W.A. and Marglin, S.A. (1969), Economics of Project Formulation and Evaluation, UNIDO, Vienna.

Chapter 11

ENVIRONMENTAL QUALITY IN A NEW SYSTEM OF SOCIAL ACCOUNTS

K.A. Fox

"People live out their lives in a sequence of environmental units; experience in these settings _is_ life. If the quality of experience is good, life expands; if it is bad, life diminishes." P.V. Gump (1971, p. 134).

Prologue

My work on social accounts began in 1973 as a response to the social indicators movement in the United States. This movement had been precipitated and encouraged by the many new social programs initiated in the Kennedy and Johnson administrations (1961-68) in an effort to create, in Johnson's words, "a Great Society".

Several prominent sociologists pointed out that the existing data systems were totally inadequate to measure progress, if any, toward the stated program objectives. They proposed legislation called "The Full Opportunity and Social Accounting Act of 1967" which would have created a Council of Social Advisers and mandated an annual Social Report of the President to Congress. One of its advocates (Bertram Gross) had been a staff member of the President's Council of Economic Advisers and another (Philip Hauser) had been an official in the Census Bureau. They evidently hoped that the proposed Council would spearhead the rapid development of social data systems of high quality embodying the advice of top university scientists in the relevant fields.

The legislation was not enacted or even debated; a conservative President (Nixon) took over in January 1969 and work on social indicators in federal statistical agencies received low priority. However, a demand for "social indicators" had been created and various private organizations and consultants tried to satisfy it in highly eclectic, ad hoc, and atheoretical ways.

I had no contact with the social indicators movement of the 1960s. However, for reasons of my own I was reading widely across the social and behavioural sciences. This interest dated back to 1930 when the first volumes of the "old" **Encyclopaedia of the Social Sciences** (1930-35) began arriving in my father's home. It intensified after 1960 under the influence of books by and/or contacts with Herbert Simon, Kenneth Boulding, Walter Isard, Richard L. Meier, the quantitative geographers William L. Garrison and Brian J.L. Berry, and the urban-regional economist and accountant Charles L. Leven. It was further stimulated by my service on the Board of Directors of the Social Science Research Council during 1963-67 and publication of the "new" **International Encyclopedia of the Social Sciences** (IESS) in 1968. My research in community development, urban-regional economics, and university planning (resource allocation in non-market systems) regularly crossed disciplinary lines. I aspired to become a well-rounded social scientist.

In 1972 I resigned my administrative position at Iowa State University and had fourteen months free for research and writing. I spent this time in Berkeley and made good use of the University of California libraries. In March 1973 the first issue of the **Social Indicators Newsletter** (Volume 1, Number 1) announced the establishment of a Center for the Coordination of Social Indicators Research under the joint sponsorship of the Social Science Research Council and the National Science Foundation. The Center was guaranteed funding for five years with the possibility of extension for another five.

The social indicators movement was alive and well! I felt I could make an important contribution to it. In September 1973 I completed a book manuscript entitled **Social Indicators and Social Theory: Elements of an Operational System**; it was published in the following year (Fox, 1974).

Despite its title, the book had a strong accounting emphasis. I pointed out that the economic indicators developed in the 1920s made little sense until Simon Kuznets and others drew a boundary around the economy as a whole and summarized all transactions within that boundary in a set of national income and product accounts. These accounts provided a common framework for the macroeconomic theories of Keynes and others, the input-output matrices of Leontief, and the dynamic econometric models of Tinbergen. Knowledge relevant to economic policy cumulated rapidly and economic data systems were expanded and improved in a coherent manner. In the United States, the creation of the Council of Economic Advisers in 1946 reflected widespread confidence that the knowledge and data had reached levels at which economists could make major contributions to national policies.

Similarly, social indicators would make little sense until we drew a boundary around the society as a whole and summarized all transactions within that boundary in a set of social system accounts. The boundary would have to include the complete living-time of all members of the society in any given year and (hence) all transactions among them. The content of the accounts should be acceptable to social and behavioral scientists in all disciplines.

The contributions of an individual to the social system might be described by a vector of initially incommensurable inputs; the rewards received by an individual from the social system might be described by a vector of initially incommensurable outputs. Social scientists from different disciplines would be challenged to define and measure these inputs and outputs in appropriate units. Purely terminological differences would presumably be recognized and a common vocabulary agreed upon for the specific purposes of social system accounting.

If members of a particular discipline chose to disregard certain categories of inputs and outputs or certain categories of time-use, they could suppress the other components of a complete system. But this would be a deliberate act; they would have to discard rows, columns, and complete matrices of numbers which were as impressive visually as the ones they kept. The existence of such accounts would encourage interdisciplinary communication and collaboration on models of social systems relevant to the analysis of policies internal to the society. Though I did not mention it specifically, the accounts and models could be opened up to reflect interactions with the physical environment.

In a 1973 paper (reprinted in Fox 1974), Paul van Moeseke and I had suggested a possible scalarization of the vector of inputs made by an individual to the social system. Our model led to a measure of the individual's **total income** as the sum of dollar values imputed to each input when the individual achieved an optimal allocation of his time according to his own preferences. On certain assumptions commonly made by economists, the equivalent dollar values of total incomes could be aggregated over individuals to yield estimates of total income for families, communities, regions, and nations.

Estimated effects on the level and distribution of total income could then be used as **one** common denominator in choosing between alternative socioeconomic programs; other criteria might, of course, contradict and override this one in particular situations. This formulation could also be used as **one** basis for evaluating interactions with the physical environment.

In the preface to **Social Indicators and Social Theory**, I expressed the hope "that this book will facilitate convergence toward an operational system or systems of social accounts and indicators.

What is called for, I believe, is an integration of theory, methods, and data across the "social" parts of the social and behavioral sciences..." (p. vii).

I did not claim to have accomplished anything of this magnitude. I did say that I had tried to read quite widely in the social sciences and to achieve a fairly uniform understanding of them at the level of the **International Encyclopedia of the Social Sciences.** "In this book I have tried to juxtapose concepts from several disciplines in a rough and preliminary fashion: time budgets, behavior settings, generalized media of social exchange, total income, transactional analysis, life cycle stages, social readjusment rating scales, functional economic areas, occupational structures, and others..." (p. x).

My chapter headings included social science concepts relevant to a system of social accounts; derivation and implications of a scalar measure of total income (written jointly with Paul van Moeseke); some other attempts to combine economic and noneconomic variables in theoretical models; time budgets, behavior settings, and total income; individuals and families; national goals accounting and policy models; social indicators and models for cities and regions; accounts and indicators for the higher education sector; occupations and earnings; elements of an operational system: I. individuals, families, and organizations in a small community, II. cities and regions, and III. national and world models and data; and social indicators and social theory. The text referred to 250 publications and the appendix (some comments on related literatures) to 250 more.

Most reviews were enthusiastic about the book. The National Science Foundation was sufficiently impressed to support my research for another six years under the title "Measurement and Valuation of Social System Outcomes". Roger G. Barker's concept of **behavior settings** proved central to my approach. Behavior settings constitute the immediate environment of human behavior and the loci of interactions between people and the physical environment - which I take to be the main theme of this book.

Behavior Settings as Basic Units in a System of Social Accounts

Roger Barker originated the concept of behavior settings nearly 40 years ago. Trained as an experimental psychologist, he had done distinguished research (some of it jointly with Kurt Lewin) in the field of child development. In 1947 he obtained funding for a long-term study of children's behavior in the settings of everyday life.

Barker established his Midwest Psychological Field Station in a small county seat town in northeastern Kansas which he code-named Midwest. It was several miles distant from any other town, and Barker found that Midwest's children under 12 spent 98 percent of their time within the town limits. Detailed studies of individual children showed that most of the variation in their behavior was accounted for by transitions between settings such as school classes, playgrounds, streets and sidewalks, home family meals, Sunday worship services, and the like. Children's behavior depended much more on **where** they were than on **who** they were. Each setting had a different "program" which imposed a different pattern of expectations on the children and elicited a distinctive pattern of behavior from them.

The same was true of adults as they moved from banks to grocery stores to bowling alleys or to business meetings. Their behavior was structured by these **behavior settings**. Collectively, Midwest's behavior settings constituted the environment of its residents' behavior. Barker's insight was later expressed as follows by Gump (1971, p. 134): "People live out their lives in a sequence of environmental units; experience in these settings is life. If the quality of experience is good, life expands; if it is bad, life diminishes".

Behavior Setting Surveys

Behavior settings are the basic units of organizations and communities. From the perspective of an experimental psychologist, a behavior setting is a marvellously complex entity. It imposes expectations and demands on each of its occupants according to his role in its program; minor deviations from the role-norm are countered almost instantly by pressures from other occupants and major ones lead to his ejection from the setting. Circumstances may confine a child to a few settings; if his behavior is appropriate to them it is not likely to change unless he "migrates" on a full-time or part-time basis to other settings which impose different demands. An adult confined to an institution or restricted by poverty or physical disability to a few settings will adapt his behavior to them; he may not be deprived of food, clothing, and shelter but he is deprived of a normal range of opportunities for social participation - he subsists but he does not belong to the mainstream of his society. An adult in the mainstream will choose to participate in a wider array of behavior settings which collectively define one of the normal, desirable, or acceptable life styles of his time and place.

In 1951-52, Barker conducted the first of a series of year-long behavior setting surveys. He made detailed observations of the 585 **community** behavior settings which existed or occurred during the year (stores, school classes and the like) and, along with many other data items, included estimates of the amount of time spent in each by town residents. He did not survey **family** behavior settings but he noted (Barker and Wright, p. 84) that the common varieties were home meals, home indoors, home outdoors, home bathrooms, and home festive occasions.

Barker was clearly aware that his methods could be extended to yield complete time budgets, for on pages 97-98 he said: "During the survey year we estimate that Midwesterners spent 5,130,000 hours in family settings, 1,030,658 hours in community settings, and 330,620 hours in foreign settings" (i.e., settings outside the limits of the town). His Figure 4.2 on page 98 shows the average hours per day spent in the three classes of settings by each of seven age groups, the two sexes, and three socioeconomic classes. Figure 4.2 accounts for 24 hours a day in each case. The larger numbers sum to 8784 hours during the year (366 days) for each of the town's 740 residents.

Social System Accounts

Most details of Barker's system are not relevant to our immediate concerns. The uses I have made of Barker's concepts are suggested by the subtitle and chapter headings of my 1985 book, **Social System Accounts: Linking Social and Economic Indicators through Tangible Behavior Settings.** The headings include behavior settings as a basis for social system accounts; the usefulness of behavior settings for classifying and describing human activities in a community; behavior settings and objective social indicators; the classification of behavior settings in social system accounts; the classification of roles in social system accounts; the classification of stocks of physical capital and consumer durables in social system accounts; the classification and delineation of communities and regions in social system accounts; a behavior setting approach to microanalytical simulation models at the community level (written by James R. Prescott); some broader implications of behavior settings for the social sciences; and social system accounts based on behavior settings: some next steps.

Behavior Settings as Basic Units of the Ecological Environment of Human Behavior

The behavior setting is not coextensive with any other unit in the social and behavioral sciences. It is the basic unit of the ecological environment of human behavior. As Schoggen (1983) observes, the **ecological**

environment is more than the physical environment, it is more than the social environment, it is not the psychological environment, it has structure and organization, and it is the source (through behavior setting programs) of powerful forces which determine much of the behavior of persons in its constituent settings.

The behavior of a particular person in a setting will be influenced by its physical aspects (walls, furnishings, counters, equipment, temperature, noise level), its social aspects (the numbers and kinds of people in it, its program, the goals of its prominent action patterns), and the demands it places on this person (the expectations governing his particular role). These various aspects of the setting influence the person's behavior to the extent that they are present in his **life space** (his awareness) at any given moment, but this appears to be the case with most people most of the time.

The other people in the setting are part of this person's environment, as he is of theirs; if only for this reason, the contents of the life spaces of different persons in the setting will be somewhat different. Each person's outline is (for him) the environment's "in-line"; the environment of each organism begins where that particular organism ends.

Behavior Settings and the Physical Environment

Environmental hazards, pollutants, and nuisances impinge on people in behavior settings. Presumably the amounts of various substances in the air and in fluids and solids relevant to activities in the setting can be measured and the amounts inhaled or otherwise absorbed by persons in the setting estimated. The same applies to noise, vibration, levels and variations of temperature, and the like. Odors, taste and appearance of potable water, and soot or dust on clothing and furniture to the extent that they are merely unpleasant but not hazardous to health would reduce total income currently.

The materials used in walls, partitions, furnishings, equipment and any other objects involved in a setting's program have made demands (directly or indirectly) on natural resources. Their manufacture may have contributed to air and water pollution at the time; their further processing (if any) in the setting may contribute to it currently; and their final disposal may also impact the environment. In many nonmarket settings the rewards to participants come from the act of participation (for example, in some family settings, in religious services, in social clubs, and in sports and recreations). It would be legitimate to study the effects of less costly and/or less environmentally-damaging materials on the total incomes (perceived, experienced) of participants in the same generic activity in structures and with equipment made of alternative materials.

Along this line, Bechtel (1977) points out that architects may say they are enclosing **space** but they are really enclosing **behavior.** Meier (1974) subtitles his book **Planning for an Urban World** as "designing resource-conserving cities". Bechtel asserts that architects **could** do a much better job of supporting and facilitating human behavior for a given expenditure; Meier implies that city planners and architects **could** support a given level of human behavior and well-being at a much lower resource cost. Bechtel makes explicit use of Barker's behavior setting concepts. Meier is aware of Barker's work but his treatment of cities as ecological systems is based on a very wide range of reading and experience.

In the next section we will show how behavior settings and the physical environment are reflected in the most mature expression of the social indicators movement up to this time - the OECD List of Social Indicators, first published in 1982.

Behavior Settings and Environmental Quality as Reflected in the OECD List of Social Indicators (1982)

At the beginning of this paper we described the social indicators movement in the United States, which began with great enthusiasm in the mid-1960s but lost its momentum and coherence during 1969-72. There was greater commitment and continuity in some other countries and in international organizations.

In particular, the OECD sponsored a Social Indicators Program which began in 1970 and published a **List of Social Concerns Common to Most OECD Countries** in 1973. Several years of complex developmental work ensued on designing indicators to measure these concerns. The **OECD List of Social Indicators** (1982) is the result.

The OECD List of Social Indicators (1982)

The OECD List of Social Indicators is reproduced in Table 1. It includes 33 specific indicators grouped under eight major headings: health; education and learning; employment and quality of working life; time and leisure; command over goods and services; physical environment; social environment; and personal safety. We believe all the objective indicators in the OECD list can be picked up and/or incorporated in social accounts based on behavior settings.

Table 1. The OECD List of Social Indicators

Social concern	Indicator
HEALTH	
Length of Life	- Life Expectancy
	- Perinatal Mortality Rate
Healthfulness of Life	- Short-term Disability
	- Long-term Disability
EDUCATION AND LEARNING	
Use of Educational Facilities	- Regular Education Experience
	- Adult Education
Learning	- Literacy Rate
EMPLOYMENT AND QUALITY OF WORKING LIFE	
Availability of Employment	- Unemployment Rate
	- Involuntary Part-time Work
	- Discouraged Workers
Quality of Working Life	- Average Working Hours
	- Travel Time to Work
	- Paid Annual Leave
	- Atypical Work Schedule
	- Distribution of Earnings
	- Fatal Occupational Injuries
	- Work Environment Nuisances
TIME AND LEISURE	
Use of Time	- Free Time
	- Free Time Activities
COMMAND OVER GOODS AND SERVICES	
Income	- Distribution of Income
	- Low Income
	- Material Deprivation
Wealth	- Distribution of Wealth
PHYSICAL ENVIRONMENT	
Housing Conditions	- Indoor Dwelling Space
	- Access to Outdoor Space
	- Basic Amenities
Accessibility to Services	- Proximity of Selected Services
Environmental Nuisances	Exposure to Air Pollutants
	- Exposure to Noise
SOCIAL ENVIRONMENT	
Social Attachment	- Suicide Rate
PERSONAL SAFETY	
Exposure to Risk	- Fatal Injuries
	- Serious Injuries
Perceived Threat	- Fear for Personal Safety

Source: Organization for Economic Cooperation and Development, The OECD List of Social Indicators (Paris: OECD, 1982), p. 13.

The OECD List of Social Indicators is still, quite properly and explicitly, in the tradition of social indicators rather than social accounts. Each of its 33 indicators is treated as a distinct entity. Even within groups (e.g., health), the various indicators involve different units of measure and there is no obvious way to assign "relative importance" weights to these units. Similarly, there is no obvious way to weight the relative importances of the eight major groups or social concerns. The OECD List, recommended for implementation by OECD member countries as their circumstances permit, includes "systematically selected statistical measures of individual well-being which can be influenced by social policies and community actions. Over time they provide a picture of some of the significant ways in which the societies of Member countries are changing". (p. 12).

The list is intended to be comprehensive in coverage. It includes very few perceptual or subjective indicators; the majority are objective measures verifiable in fact or in principle by trained observers and collected or collectible in official reporting systems, censuses, and sample surveys. The one clearly subjective indicator "perceived threat, or fear for personal safety" could also be collected and interpreted in terms of the time-and-place settings in which the threat is perceived to be concentrated.

The OECD List as a Vector of Outputs of Behavior Settings

Behavior settings with the same programs can be aggregated into categories analogous to industries. Barker (1968) found 884 nonhousehold behavior settings in Midwest during 1963-64 but only 198 different programs. There was only one bank but there were four gasoline service stations, three grocery stores, several age-graded classes in the elementary school, and 103 settings of **business meetings** all conducted in general accordance with Roberts' Rules of Order! Central place theory and common observation (not to mention economic censuses) indicate that these 198 programs were found in most other small towns in the United States. Additional programs were found in successively larger cities, but the total number of distinctly different programs in the United States according to Barker's criteria was probably around 400.

We may visualize a matrix with 400 columns for the different categories of behavior settings, and rows representing subgroups of the U.S. population classified by age, sex, occupation, socioeconomic status, or any other attribute or combination of attributes. The entries in each row would be the numbers of person-hours spent by members of the specified population subgroup in the various categories of settings during a given year; the row totals would include 8760 person-hours for each member. Each row would be a **time-allocation vector** (TAV); on a per-member basis it would represent the average life style of the group. The matrix as a whole would be a **time-allocation matrix** (TAM).

Now, suppose we record above the TAM a matrix with 400 columns and 33 rows, one for each of the 33 OECD indicators. What contribution does each category of behavior settings make to each indicator? What would specified changes in the attributes of these settings contribute to changes in the values of each indicator? The present indicators may not be ideally formulated to represent behavior setting outputs in all cases. Changes in the levels of some indicators depend on macroeconomic policies and cannot be influenced by changes initiated at the behavior setting level. But several of the indicators do reflect interactions with the physical environment.

OECD Indicators Relating Aspects of the Physical Environment to Individual Well-Being

Three of the OECD indicators relate to housing conditions: indoor dwelling space (number of persons per room); basic amenities (private kitchens, bathrooms, flush toilets and other features); and access to outdoor space (private or shared gardens, roof gardens, terraces, balconies, nearby parks or plazas). A fourth is based on the proximity of residents, in minutes, to medical emergency services, stores satisfying daily needs, a pre-primary school, a primary school, a sports center where pupils can participate, a post office, and a public transport stop. All these services involve physical capital - structures and equipment. Two other indicators, exposure to air pollutants and exposure to noise, are to be measured outdoors in residential areas.

All six of the OECD indicators so far discussed are measurable attributes of the physical environments in which people live; their levels can be improved only through operations in and on these environments. A seventh, travel time to work, is also a fact of the physical environment. An eighth, work environment nuisances, also involves buildings, equipment, vibration, heat, cold, dust, noise - all attributes of the physical environment.

Concluding Remarks

Barker (1986) rated each behavior setting on eleven **action patterns**: aesthetics, business, education, government, nutrition, personal appearance, physical health, professionalism, recreation, religion, and social contact. These patterns were carefully defined. They reflect the purposes for which the settings were created. They are goal-directed or output-oriented. They are the microsocial counterparts of social concerns.

Behavior settings exist in time and space. They are highly visible and tangible features of the environment of everyday life. The same settings can be directly observed by social and behavioral scientists of all disciplines, environmental scientists, engineers, architects, city planners, administrators, and public officials. They are basic units for social accounting systems and for observing interactions between human activities and the physical environment.

Some final words: My approach (following the pioneering work of Roger Barker) is called "eco-behavioral science". We move about and behave in the physical environment, but our behavior is guided by economic, social, and cultural values. Most of the relevant environment we built ourselves, for economic, social, and cultural reasons. Recent sustainability discussions have increased my confidence in the importance of eco-behavioral science as a common meeting ground for social and environmental scientists.

References

Barker, R.G., (1968), Ecological Psychology: Concepts and Methods for Studying the Environment of Human Behavior, Stanford, Cal.: Stanford University Press.

Barker, R.G. and Wright, H.F. (1955), Midwest and Its Children: The Psychological Ecology of an American Town, New York: Harper and Row. Reprinted 1971 by Archon Books, Hamden, Connecticut.

Bechtel, R.B., (1977), Enclosing Behavior, Stroudsburg, Pennsylvania: Dowden, Hutchinson and Ross.

Fox, K.A., (1974), Social Indicators and Social Theory: Elements of an Operational System, New York: John Wiley.

Fox, K.A., (1985), Social System Accounts: Linking Social and Economic Indicators through Tangible Behavior Settings, Dordrecht, Holland: D. Reidel and Boston: Kluwer Academic Publishers.

Fox, K.A. and Van Moeseke, P. (1973), Derivation and Implications of a Scalar Measure of Social Income, in: Bos, H.C., Linnemann, H. and De Wolff, P. (eds.), 1973, Economic Structure and Development: Essays in Honor of Jan Tinbergen, Amsterdam: North-Holland and New York: American Elsevier.

Gump, P.V., (1971), The Behavior Setting: A Promising Unit for Environmental Designers, Landscape Architecture, vol. 61, no. 2, pp. 130-134.

Meier, R.L., (1974), Planning for an Urban World: Designing Resource Conserving Cities, Cambridge , Mass.: M.I.T. Press.

Organization for Economic Cooperation and Development, (1982), The OECD List of Social Indicators, OECD, Paris.

Schoggen, Ph., (1983), The Ecological Environment of Molar Human Behavior, (paper presented at the 91st Annual Convention of the American Psychological Association, Anaheim, CA, August).

Chapter 12

MULTICRITERIA ANALYSIS: A DECISION SUPPORT SYSTEM FOR SUSTAINABLE ENVIRONMENTAL MANAGEMENT

P. Nijkamp

Introduction

A balanced and sustainable socio-economic **and** environmental evolution of our societies appears to be more a fiction than a reality. Our economic and environmental systems seem to be controlled by antagonistic forces, so that current quantitative and qualitative changes in our demographic, industrial, social and technological patterns seem to exhibit incongruent economic and environmental repercussions. Thus, conflicts seem to be omnipresent in the area of environmental and resource management, and in policies aiming at an ecologically sustainable economic development.

In this context, various types of conflicts may be distinguished:

- **inter-actor conflicts**; examples of such conflicts are differences in priorities attached to economic growth vis-à-vis environmental quality by **various groups** in society (e.g., industrialists versus environmentalists) or **different decision-making bodies** (e.g., a ministry of transportation infrastructure versus a local city council).
- **inter-regional conflicts**; examples of such conflicts are the geographical transfer of negative externalities (notably, waste) to **neighbouring regions** (for instance, the Rhine pollution) or to the **earth as a whole** (for instance, ocean pollution, ozonization, acid rain etc.).
- **inter-temporal conflicts**; examples of such conflicts are **backward** conflicts emerging from the preservation of our cultural and natural heritage (e.g., monuments) and **forward** conflicts emerging from the present use of scarce resources which may be detrimental to the interest of future generations (e.g., depletion of fairly scarce natural resources).
- **intra-person** conflicts; such conflicts emanate as a result of contradictory interests **within** one decision-maker (e.g., a person with an environment-conscious constituency who has a conspicuous consumption pattern). One may even argue that the latter type of conflict shapes the conditions for the emergence of the other conflicts mentioned

above.

In view of the above mentioned conflicts, it is evident that the idea of a sustainable development (cf. the Brundtland Report) cannot easily be attained. On the other hand, it is noteworthy that the notion of conflicts does by no means imply that always a **zero-one decision situation** in the case of environmental economic problems exists. In many cases, decision situations are of the **incremental** type for which, for instance, some reduction in pollution has to be traded off against some increase in unemployment. Furthermore, there are even many situations in which environmental objectives reinforce economic objectives, witness notions like 'nature building', 'steady state growth', 'eco-stability' etc.

The intertwined nature of all processes in an economic-environmental system calls for an adequate methodology that is able to reconcile - to some extent and in principle - economic and ecological paradigms from a steady state and/or longterm perspective (see Nijkamp and Soeteman, 1988, and Repetto, 1986). Traditional economic approaches - such as marginal opportunity cost principles or shadow cost approaches in alternative accounting systems - neglect usually the intriguing problem of environmental externalities and, in any case, the qualitative shifts in dynamic economic-ecological systems. Besides, recent advances in the field of irreversible externalities (based inter alia on flexible discount rates for multiregional choice problems) often lack operationality, so that the actual causes of natural resource degradation cannot be coped with effectively in a conventional market system.

Clearly, a major problem consists of the local scale of environmental externalities, in terms of both causes and effects. For example, global problems such as acid rain, sedimentation, desertification, ozonization, eutrofication, ocean pollution and resource extraction are often the result of a great many small-scale and local actions (without an environment watchful constituency), while also the impacts can be observed most clearly at a local or regional scale. This problem of spatial openness can unfortunately hardly be incorporated in conventional analytical tools.

In addition, the pluriform nature of **dynamic** ecological and economic processes cannot be described either meaningfully by means of conventional measurement models adopted in an isolated monodisciplinary framework, especially because data limitations hamper the use of a rigorous analytical framework (cf. Brouwer, 1987).

However, integrated models may potentially be powerful instruments to analyze and to evaluate changes in the (a-)biotic environment caused by human activities such as consumption and production of goods and services, recreation, or environmental protection measures. Initial attempts in this area can be found in the generalized (economic-ecological) input-output approaches developed among others by Isard and Leontief in the seventies. These models were static in nature and did not aim at depicting simultaneously a set of heterogeneous components of change processes, but served to provide a consistent accounting system (based on static equilibrium concepts). Apart from the static nature, the first generation of integrated environmental models was neither particularly appropriate for the analysis of concrete environmental policy issues such as land use planning, land reclamation, water resource planning, recreational planning or urban renewal.

From the late seventies onwards, various new attempts have been made to develop operational integrated environmental models in various fields of resource, land use and infrastructure planning. The interest in such analytical tools emerged from the growing concern among local and regional authorities about environmental quality. Furthermore, recent advances in the field of integrated systems modelling (based on computer sciences, e.g.) also led to an increase in the quality of suitable models for multidisciplinary purposes. Despite this progress, however, it is still true that the popularity of monodisciplinary environmental models (based on either economic or ecological principles, e.g.) is much higher than that of fully integrated environmental models. Attempts at integrating different model types are thus far still unsatisfactory (cf. Jansson, 1984, or Kemp and Long, 1984).

In the light of the above mentioned observations, the need for a coherent analytical policy framework in which economic goals can be traded off properly against environmental interests is clear. In the next section, the backgrounds of such a more appropriate methodology will be outlined.

A Multidimensional Analysis Framework for Balanced Environmental Management

The various types of conflicts described in section 1 call for an adequate methodological framework which may shape the basis for a balanced environmental management. This brings us into the realm of evaluation studies.

Policy analysis or evaluation, in general, aims at rationalizing planning and decision problems by systematically structuring all relevant aspects of policy choices (for instance, the assessment of impacts of alternative choice possibilities). Evaluation is usually not a one-shot activity, but takes place in all phases of decision-making (for instance, on the basis of learning principles). In addition, a systematic support to complex planning and decision problems presupposes a balanced treatment of too many details and too little information. Besides, the results of an evaluation procedure have to be transferred to policy-makers in a manageable and communicable form, particularly because the items of an evaluation problem are usually multidimensional in nature (including incommensurable or even intangible aspects). Finally, it has to be realized that the "planning environment" is usually highly dynamic, so that judgements regarding the policital relevance of items, alternatives or impacts, may exhibit sudden changes, hence requiring policy analysis to be flexible and adaptive in nature. Rigid evaluation techniques run the risk that an evaluation does not cover all planning issues in a satisfactory way, so that biased decisions may be taken.

There is considerable agreement among planners and decision-makers that evaluation methods have considerable benefit in making values and choices more explicit. However, there is no single evaluation method which is able to deal consistently with the wide variety of contexts and contents of public planning and decision-making. However, any evaluation requires appropriate information. The aims of the evaluation, however, may be different and depend on actual institutional and administrative interest. Three broad categories of behavioural paradigms may be distinguished for public decision-making:

- "optimizing" behaviour (the conventional economic paradigm);
- "sasisficing" behaviour (the behavioural decision-maker's paradigm);
- "justificing" behaviour (the pragmatic policy-maker's paradigm).

Although the majority of formal evaluation techniques is focusing attention on the first category and to a large extent also on the second category, in policy practice evaluation is often used as a means of justifying policy decisions, even if the actual decisions are not in agreement with "optimizing" or "satisficing" principles. In any case, however, relevant data for a policy judgement have to be collected. Such data should be represented in an operational form in order to make the actual choice issues as transparent as possible.

Clearly, every planning or policy-making process presupposes choices to be made, regardless the evaluation methods for planning or policy-making that have been adopted. The difficult task of choosing among alternative courses of action (choice possibilities, options etc.) cannot be circumnavigated and is common to all public planning and administration problems.

Beside the (institutional or administrative) structure of a decision problem, the specific plan evaluation method to be used will also determine which data are requested for the policy analysis at hand. For instance, checklist approaches, cost-benefit sudies, planning balance sheet techniques, goals achievement methods, multiple criteria analyses, multiple objective programming models have all their own specific data needs. Clearly, assuming a cyclical model of planning implies also that the relevance of an evaluation technique has to be judged in light of the available data. In this regard, it is worth noting that monitoring is a necessary ingredient of an adaptive evaluation methodology, so that in each phase of the planning problem both the data and the evaluation method can be critically judged.

Evaluation may relate to both **sectoral** planning issues (transportation network planning, facilities planning, e.g.) and **integrated** planning issues (comprehensive regional or urban planning, e.g.). In all cases, there should be a close agreement between available information and evaluation.

Information plays a crucial role in any evaluation method. Two important aspects of information are worth mentioning in the present context.

First, any policy decision will affect the welfare position of individuals, regions or groups in a different way, via either pecuniary or non-pecuniary external effects. Consequently, the public support for a certain policy decision will very much depend on the distributional and external effects of such a decision. Thus, in general, there is a need for evaluation methods that try to assess the pros and cons of a certain choice alternative for separate groups or regions. Information on such gains and losses are not always cardinal in nature; also qualitative, fuzzy or verbal information may provide a meaningful input for a policy analysis. Altogether, spatial and/or social referencing of information is highly desirable to make evaluation more effective and succesful. It goes without saying that this notion of distributional and external effects is of particular relevance in the context of environmental management which is usually dealing with conflicting options between different groups or actors.

In the second place it should be noted that information has a much deeper meaning than just a set of data. **Data** are only numerical representations of attributes of people, organizations, objects or event. **Information** may be defined as a collection of organized data (for instance, by means of statistical techniques, modelling or transformation) so as to provide structured and systematic insight regarding a phenomenon. In this regard, an information system means any kind of systematic and coherent analytic or decision support system for planners and policy makers. Such a system serves to contribute to solving, organizing or rationalizing complex choice and decision problems. Thus structured information is a vehicle for communication between all actors involved. If one adopts a procedural view of planning, in which decision-making is regarded as a process, information systems have to be flexible, so as to provide at any desired moment decision agencies with specific tailormade information. This procedural view of planning leads thus in essence to the design of **adaptive** information system, which have gained much popularity in recent years. This tendency runs parallel to the recent design of interactive user-oriented multiple criteria decision models. Such models may be regarded as decision support systems and have also a great potential in the field of environmental management. The next section will be devoted to a brief introduction to multiple criteria evaluation methods.

An Introduction to Multiple Criteria Evaluation

Policy evaluation has become a major component of modern public planning and administration. The history of plan and project evaluation, however, shows that various phases of evaluation analysis can be distinguished.

The early history of plan and project evaluation (before World War II) was marked a strong tendency toward a financial trade-off analysis. A good example can be found in the USA, where a specific emphasis was placed on the regional growth aspects of an integrated regional hydrological project in the Tennessee Valley.

Later on much attention was focused on the judgement of alternative American military defense systems, based **inter alia** on cost-effectiveness principles. After World War II social cost-benefit analysis gained increasingly popularity in public policy evaluation. The cost-benefit methodology became a leading evaluation instrument until the seventies, especially in countries with a market economy or a mixed economy. Several limitation inherent in cost-benefit analysis were relaxed by introducing amendments such as planning-balance methods, goals-achievement methods or shadow project methods.

From the seventies onwards a new class of evaluation methods emanated called multiple criteria analysis. Many origins of the class of multiple criteria methods can be found in France (Roy, Guigou, Jacquet-Lagrèze, e.g.). Especially the well-known ELECTRE-techniques (or concordance techniques) have become a mainstream in current evaluation methodology in many countries and form many areas of policy research.

The following reasons can be mentioned for the increasing influence of multiple criteria evaluation techniques vis-à-vis conventional evaluation tools:

- the impossibility of including intangible and incommensurable effects in the conventional evaluation techniques;
- the conflict nature of modern planning problems, so that - instead of a single decision-maker - various (multi-level) formal and informal decision agencies determine a final choice;
- the shift from conventional "one-shot" decision-taking to institutional and procedural decision-making where many political aspects play a major role;
- the desire in modern public decision analysis **not** to be confronted with a single and "forced" solution but with a spectrum of feasible solutions.

All these reasons have led to the current popularity of multiple criteria analysis in public planning. In the seventies and the beginning of the eighties a real avalanche of multiple criteria methods has taken place, so that at the moment there is a wide array of various multiple criteria evaluation methods for public planning and evaluation. These methods can be used for different purposes and in different contexts. The following distinctions can be made regarding the contents and scope of multiple criteria evaluation methods.

(1) **discrete versus continuous methods**

Discrete evaluation methods focus attention on a finite wet of (a priori known) choice alternatives, whereas continuous evaluation methods are - in principle - related to a non-countable (and hence not precisely identifiable) set of choice alternatives.

(2) **multi-person versus single-person evaluation**

In case of multi-person (or multi-committee) evaluation problems, it is in general impossible to assume unambiguous and a priori known trade-offs, so that flexibility allowing for dynamic preference articulation and bargaining procedures has to be ensured. For a single-person case it is often easier to specify policy priorities.

(3) **identification versus selection of alternatives**

In various evaluation problems it is only necessary to identify a limited set of reasonable (or 'satisficing') choice possibilities, whereas in other cases the demands are put much higher, viz. the unambiguous selection of a single alternative. In the first case, it may be sufficient to find a set of non-dominated (or Pareto) solutions for which the value of one policy objective cannot be improved without reducing the value of competing criteria.

(4) **single-step versus multi-step evaluation procedures**

Single-step evaluation takes for granted that a given evaluation problem must be solved immediately, whereas multi-step evaluation assumes a process character for evaluation (e.g., learning mechanisms, adaptive information provision).

(5) **soft versus hard information**

Soft evaluation problems are characterized by non-metric information (e.g., ordinal data, qualitative statements), whereas hard problems are based on quantitative (e.g., cardinal) information. An intermediate case is **mixed** information, which includes both qualitative and quantitative information.

Reviews of various types of multiple criteria evaluation methods can be found among others in Isard and Smith (1982), Nijkamp (1980), Rietveld (1982), and Voogd (1983). In all cases, multiple criteria methods are always marked by two types of information, viz. the **effect score matrix** (i.e., the numerical assessment of all relevant impacts of a set of choice alternatives each of them being measured in its own dimension) and a **preference or weight vector** (i.e., the numerical assessment of the relative priority attached to each of the decision criteria considered in the effect score matrix). A wide spectrum of techniques may then serve to find a relevant answer, depending on the specific nature of the information used and on the scope and contents of the evaluation concerned. This problem will be discussed in section 4.

Choice of Method

At present, at least 40 to 50 different methods can be distinguished. This raises of course the question which methods are superior or inferior, or more precisely: **which** methods are suitable for **which** class of evaluation problems? This will be the major aim of the present section. Despite the large number of (simple and sophisticated) methods that are currently available for plan and project evaluation, there is still surprisingly little information available regarding the specific features of evaluation methods and the precise conditions under which a method is chosen (for instance, by coincidence because an analyst has only become familiar with this method). It is quite common that the evaluation problem is reformulated in order to let it comply with the features of the evaluation method at hand (see Figure 1a).

Therefore, an alternative methodology will be presented here. A central part of the paper is made up by an attempt at systematically designing a typology of classes of specific evaluation problems (characterized by multiple attributes), so that then the requirements and criteria can be specified which have to be fulfilled by any evaluation method to be used in this context. This approach provides a logical framework for consistently eliminating and selecting evaluation methods for specific classes of planning and management problems (see Figure 1b.).

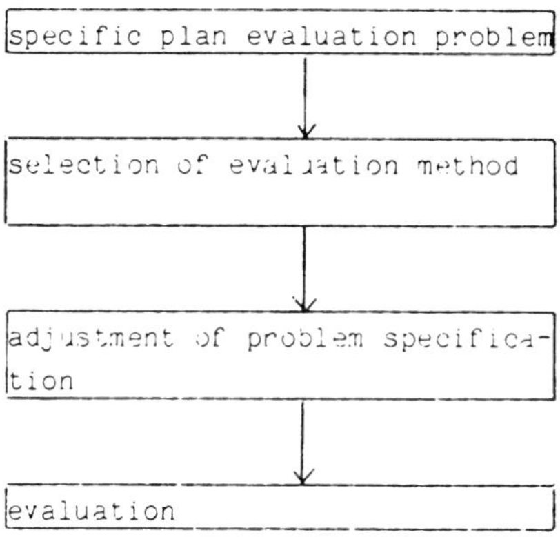

Figure 1a. Traditional phases in policy evaluation

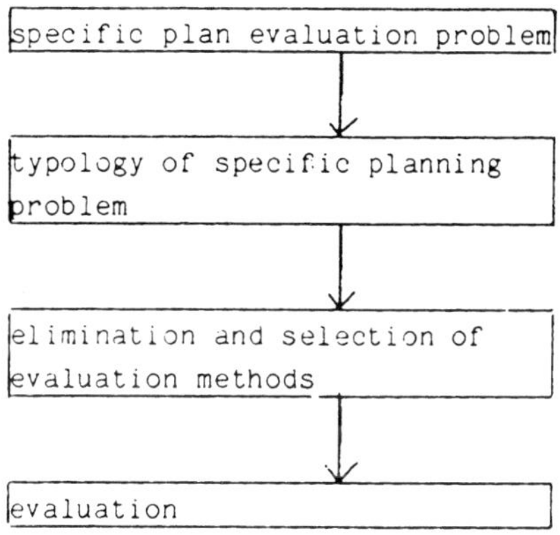

Figure 1b. Methodology adopted in this paper.

The second approach is more flexible, as it prevents a researcher from an unjustifiable reduction or reformulation of an evaluation problem in order to let it pass through the narrow filter of an ad hoc selected evaluation method. In essence, the problem discussed in this section is a meta-multiple criteria problem: how to eliminate and/or choose a certain evaluation method for a given policy problem, while taking into account a set of multiple methodological judgement criteria?

In order to provide a concrete answer to this intriguing question, an important area from public planning has been selected, viz. environmental planning and management, in order to design a methodology for the choice of decision problems. These environmental planning and management issues have in general close links with regional and urban planning, physical planning, energy planning, and transportation planning, so that a whole range of evaluation problems in a spatial context can be covered by this methodology.

A closer examination of this question has to be based on a consistent inventory and classification of environmental evaluation problems, and guided by a set of relevant classification principles. Therefore, we have to focus attention on the following precise research question: is it possible to specify a set of discriminating criteria for classes of environmental planning and management problems in order to be able to select an appropriate set of evaluation methods for a specific class of environmental planning and management problems?

To answer this question the following steps have been undertaken:

- inventory and classification of environmental management problems according to a set of policy relevant attributes;
- identification of criteria for selecting an evaluation method in relation to the relevant attributes;
- confrontation of the selection criteria with the features of available evaluation methods.

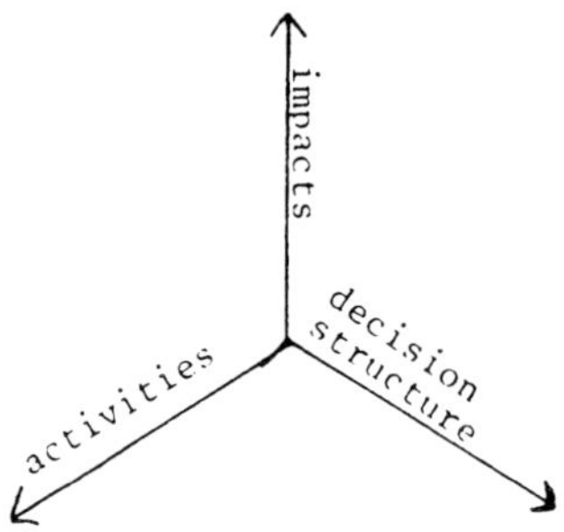

Figure 2. Three dimensions of environmental evaluation problems.

This problem is essentially a multidimensional classification problem. In order to characterize a given environmental evaluation problem, three main categories of features of such a problem may be used, viz.

(1) The attributes of the activity leading to the environmental problem in question (i.e., the activity profile).
(2) The characteristics of the impacts caused by the above mentioned activity (i.e., the impact profile).
(3) The nature of the decision structure (or institutional structure) related to the activity concerned (i.e., the decision profile).

The three profiles related to these categories will be sketched here briefly.

(1) **activity profile**

Activities causing a significant environmental impact may be represented among others by means of the following categories:

- transportation (infrastructure)
- land use
- resource use
- landscape planning
- water management
- energy use
- industrial production
- etc.

Each of these categories can be further typified according to the question whether they are plans, projects or regulations.

(2) **impact profile**

The impacts of an activity may relate to both environmental impacts (pollution, noise nuisance etc.) and non-environmental impacts (employment, income, etc.). Clearly these impacts have also various other dimensions, like geographical scale, temporal scale, and applicability of formal regulatory rules.

(3) **decision profile**

The decision profile describes among others the number of alternatives to be taken into account, the information content of each alternative, the institutional decision structure for each of the alternatives, and the aim of the evaluation itself (e.g., selection, communication, etc.).

Next, by means of the previous schematic typology, each specific environmental evaluation problem can be characterized through a complete list of features, the so-called demand profile.

In the following step the set of available evaluation methods is further examined. A first screening leads then already to an elimination of various evaluation methods. For example, in case of a discrete evaluation problem, the set of continuous methods can be neglected. Similarly, one may use the level of available information (e.g., qualitative information) to further reduce the set of methods to be considered.

After the elimination phase, each of the remaining methods is characterized by examining whether or not this specific method is able to

take into account the criteria and features mentioned in the "demand" profile (the specification of the "supply" profile).
Finally, by comparing the "demand" profile with each of the "supply" profiles of the corresponding evaluation method, one may select one or more methods which assure a maximum correspondence between "supply" and "demand" profile.

This procedure has been applied fruitfully in several cases (see Janssen et al., 1984, and Janssen and Nijkamp, 1986) and ensures a consistent methodology for the choice of an appropriate evaluation method.

After our discussion of the context within which multiple criteria evaluation can be used, it is now evident that environmental management problems lend themselves in particular for a broad judgement on the basis of multiple criteria methods. Its potential as a decision support system enhances even its value. In the next section some empirical illustrations on the use of such methods will be given.

Empirical Illustrations of the Use of Multiple Criteria Evaluation for Environmental Management

In the present section two illustrations of decision support tools for environmental management will be given. The first relates to an integrated national-regional environmental-economic model for the Netherlands, based on a study by Hafkamp (1984), while the second one relates to an environmental decision support system for pollution control in the Netherlands, based on a study by Janssen and Nijkamp (1988).

(1) A triple layer model for environmental management

In this subsection, a brief description of some results reached with a Triple Layer Model (TLM) for the Netherlands will be given. This is a model focussing on environmental impacts of economic development in the Netherlands. The core of TLM is a large non-linear national economic model, specified in terms of four main economic sectors. This model, which has been developed before TLM came into being consists of modules for production, final demand, imports, wages, prices, incomes, social insurance, employment, etc. TLM can be characterized to give a breakdown of results produced by the national model for five regions. This regional breakdown is accomplished for three variables: production, income and employment. The breakdown is not a top-down approach via a fixed share assumption, but includes detailed regional information in the form of regional input-output tables.

TLM also contains an environmental model to generate regional emissions of various pollutants on the basis of regional production values. Thus, the model can be used to generate the regional consequences in terms of income, employment and pollution as a consequence of national economic developments.

The interactive features of TLM have been introduced to study the conflicts between various policy objectives playing a role in the model. Conflicts exist between objectives from different sectors (e.g. environment versus income), but also between objectives from different regions (region A versus region B). The interactive features of TLM can be used by decision makers to generate policy alternatives in a stepwise way, which are in agreement with his priorities. A stepwise approach is chosen, since it is usually very difficult for decisionmakers to formulate priorities at once when several conflicting objectives are involved.

In every step of the interaction, TLM is used to generate a small number of alternatives (see Table 1 as an example). Three extreme alternatives are obtained by maximizing the main objectives at the national level (income, employment, environmental quality). In addition, a compromise alternative is determined which produces outcomes which are as near as possible to the maximum attainable levels for the three alternatives (see Hafkamp, 1984 for an operationalization of this notion). The effects of each alternative on the 3 objectives at the national and regional level can be found in Table 1.

As a response to this information, the decisionmaker may express his opinion on the compromise alternative by indicating which objective(s) for which region(s) are least attractive. For example, if environmental quality at the national level is indicated as the objective with the least favourable performance, this is interpreted as a side constraint added to the TLM model, saying that in the next step environmental pollution must be less than 74.900.

Thus, after the decision maker's response, TLM produces in a next step a new selection of alternatives which better reflect his priorities. It can be shown that this interactive procedure leads to convergence: it enables the decisionmaker to select the alternative which best reflects his priorities.

Table 1
Results of the interactive optimization procedure. Iteration no. 1.

Variable	Optimization of Income (10^9 Dfl/y)	Employment (10^6 man y.)	Environm. quality	First compromise solution
National income	134.600	133.300	130.400	134.200
National employment	3.360	3.390	3.330	3.380
Environmental qual.	77.000	76.200	72.700	74.900
Income, Region 1	14.200	14.100	12.600	14.200
Income, Region 2	22.300	21.400	20.400	21.900
Income, Region 3	29.000	29.500	28.600	28.600
Income, Region 4	32.600	32.100	34.100	32.800
Income, Region 5	36.500	36.200	34.800	36.700
Employment, Region 1	0.309	0.313	0.306	0.312
Employment, Region 2	0.582	0.587	0.575	0.585
Employment, Region 3	0.827	0.833	0.818	0.831
Employment, Region 4	0.775	0.782	0.767	0.781
Employment, Region 5	0.869	0.874	0.859	0.873
Environm. qual. 1	6.460	6.570	6.410	6.240
Environm. qual. 2	10.720	9.630	9.780	9.630
Environm. qual. 3	23.110	25.830	23.300	22.240
Environm. qual. 4	16.230	15.270	14.120	16.070
Environm. qual. 5	20.530	18.890	19.130	20.540

(2) A decision support system for environmental management

In this subsection an interactive model will be presented which may serve as a multicriteria decision support system for environmental management for the Netherlands. This Environmental Assessment System (EASY) is a decision support system which assists decision-makers in the assessment and evaluation of alternative strategies in electricity production. This sytem can be used to compare different combinations of conversion systems (e.g., combinations of coal, nuclear power, and oil) regarding their environmental effects (in terms of their contribution to acidification, deforestation etc.). The decision on the best electricity generation mix can be described as semi-structured. Even when there would be a consensus on the type of impact method to be used, the measurement of information, the choice of alternative options, and the type of appraisal criteria, there is in general no consensus on the exact specification of choice options, the damage categories to be considered, the level of threshold values for damage, the relative importance attached to the various categories, etc.

The main features of EASY will be described here in relation to problem formulation, impact assessment and evaluation (see also Figure 3). In EASY, the problem formulation concerns the identification of alternatives, attributes of the alternatives and appraisal criteria. The alternatives are formed by including individual conversion units, such as a coal unit, a nuclear unit, etc., into different combinations adding up to a specified production capacity. Attributes of the alternatives include location, emissions and land use. Examples of appraisal criteria are: forest damage, health damage and agricultural damage.

Impact analysis in EASY deals with effect scores, preferably in combination with the distribution of the effects in time and space. Environmental quality is calculated for 20 regions in the Netherlands and for 5 specified time points. Environmental quality includes concentrations, percentiles and depositions of SO_2 and NO_x, waste production, land use, etc.

Fig. 3. Functions of an environmental DSS.

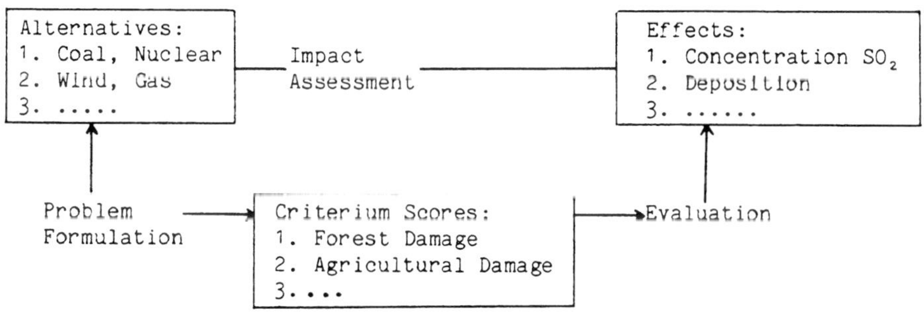

Evaluation results in an appraisal table and/or in a ranking of the alternatives. In EASY appraisal scores are calculated by relating effect scores to effect-thresholds and damage categories. The ranking of the alternatives is based on an assessment of the relative importance of damage categories.

To perform these three functions EASY contains three main elements: the data base (including attribute values of relevant conversion units, features of receptor areas, effect thresholds etc.) controlled by the data base management system (DBMS); and the user interface (including e.g. visual presentation results).

At present EASY contains a dispersonal model to predict environmental quality, a multicriteria method and various aggregation procedures. Dispersion is calculated through the use of source-receptor matrices. This is computationally simple and accurate enough for the purposes of EASY, viz. to compare different choice alternatives. The Model Base Management System selects the appropriate model, requests the relevant user instructions and data, and supplies the results.

Impact analysis results usually in a relatively large amount of diverse information. Since the human mind can only handle a limited amount of information simultaneously, it is usually impossible to make an appropriate comparison of alternatives. The aim of evaluation is to structure, condense and present the information in a way that facilitates comparison of choice alternatives.

EASY generates a 3-dimensional effect matrix containing 5 alternatives, 40 effects (effects include concentrations, depositions and land use) and 20 regions. Part of this matrix is illustrated in tables 2 and 3.

The results of these models can also be used in adaptive decision support systems by varying the extent of policy control. Furthermore, the model is also able to present results in graphical form, either as diagrams or as computerized map presentations for various regions in the Netherlands.

The above mentioned illustrations have shown the power and potential of interactive decision support models based on multiple criteria evaluation. In this context it is clear that a balanced and sustainable environmental management may be supported strongly by the use of modern evaluation tools.

Regions	Background levels	ALT 1	ALT 2	ALT 3	ALT 4	ALT 5
Gron.	54.8	81.1	55.4	55.6	54.9	54.8
Friesl.	43.5	65.1	43.9	44.0	43.6	43.5
Drenthe	53.4	69.2	53.9	54.2	53.5	53.4
K. v. Over.	54.4	69.8	54.8	54.9	54.5	54.4
Twente	60.4	69.1	60.7	60.7	60.5	60.4
Veluwe	58.6	70.5	58.8	58.9	58.7	58.6
Achterh.	72.4	81.3	72.8	72.7	72.5	72.4
Betuwe	62.3	71.6	62.6	62.7	62.5	62.3
Utrecht	59.2	72.4	59.4	59.5	59.4	59.2
K. v. N.H.	59.2	83.3	59.8	59.6	59.3	59.2
Zaanstr.	67.0	90.0	67.2	67.4	67.2	67.0
Leiden	82.6	102.3	82.8	83.0	82.9	82.6
Rijnmond	65.0	81.2	65.1	65.5	65.3	65.0

Table 2. Alternatives and Regions for Effect: Concentration NO_x

Pollutants	Threshold levels	Background levels	ALT 1	ALT 2	ALT 3	ALT 4
SO_2		27.7	41.8	27.2	28.0	27.9
SO_2P50	75.0	20.5	30.9	20.5	20.7	20.7
SO_2P95	200.0	77.2	116.6	77.4	78.1	77.9
SO_2P98	250.0	100.7	152.1	100.9	101.8	101.7
SO_2DD		1182.1	1138.2	1184.8	1198.7	1195.2
SO_2ND		403.2	585.3	403.8	410.5	407.1
SO_2TD		1585.2	2423.5	1588.6	1609.2	1602.3
SO_4		7.7	9.1	7.7	7.7	7.7
NO_x		28.1	39.3	28.2	28.5	28.3
NO_2		30.5	38.1	30.6	30.7	30.6
NO_2P50		28.1	35.1	28.1	28.3	28.2
NO_2P98		78.1	97.5	78.2	78.7	78.4
NO_2P98	135.0	65.0	81.2	65.1	65.5	65.3

Table 3. Alternatives and Effects for Region: Rijnmond

References

Brouwer, F. (1987), Integrated Environmental Economic Modelling, Martinus Nijhoff, Dordrecht.

Bullinger, H.J. et al. (eds.) (1988), Information Management for Organisational Systems, Elsevier, Amsterdam, pp. 917-925.

Hafkamp, W. (1984), Economic-Environmental Modelling in a Nationaln-Regional System, North-Holland Publ. Co., Amsterdam.

Isard, W., and Smith, C. (1982), Conflict Analysis and Practical Conflict Management Procedures, Ballinga, Cambridge.

Janssen, R., and Nijkamp, P. (1988), Interactive Multicriteria Decision Support for Environmental Management, Paper Eurinfo 88 Conference, Athens, May 1988.

Janssen, R., and Nijkamp, P. (1986), A Typological Approach to Multiple Criteria Conflict Analysis, Conflict Management and Peace Science, vol. 9, no. 2, pp. 45-53.

Janssen, R., Nijkamp, P. and Voog, H. (1984), Environmental Policy Analysis: Which Method for Which Problem?, Revue d'Economie Régionale et Urbaine, vol. 14, no. 5, pp. 839-855.

Jansson, A.M. (ed.) (1984), Integration of Economy and Ecology, Wallenberg Foundation, Stockholm.

Kemp, M.C. and Long, N.V. (eds.) (1984), Essays in the Economics of Exhaustible Resources, North-Holland Publ. Co., Amsterdam.

Nijkamp, P., and Soeteman, F. (1988), Environmental Policy between Economic Development and Ecological Sustainability, International Journal of Social Economics, vol. 15, nrs. 3/4, pp. 88-102.

Repetto, R. (1986), World Enough and Time, Yale University Press, New Haven.

Rietveld, P. (1982), Multiple Objective Decision Methods and Regional Planning, North-Holland Publ. Co., Amsterdam.

Voogd, H. (1988), Multicriteria Evaluation for Urban and Regional Planning, Pion, London.

Chapter 13

SOCIAL COST BENEFIT ANALYSIS AND SOIL EROSION

P. Dasgupta
K.-G. Mäler

Introduction

The theory of social cost-benefit analysis, although originally much motivated by public water management problems, did not until recently take much note of investment projects in which environmental resources are a prominent item. For example, the several treatises on investment planning and social cost-benefit analysis written during the decade of the 1970s (e.g., Dasgupta et al., 1972) and Little and Mirrlees (1969, 1974) made no mention of environmental resources. While these publications spelt out in detail the social costs of employing unskilled labour, tradeable commodities, public versus private investment, and so forth, the specific problems that are associated with the valuation of environmental resources were untouched in such discussions. This omission is presumably a reflection of the fact that environmental resources, until recently, did not enter the agenda of economic debates in developing countries. It is tempting to speculate why this may have been so.

Part of the reason, we would conjure, is that discussions in the Western industrialized countries on environmental problems were initially restricted in great part to air and water pollution, and destructions of places of scenic beauty. The subject appeared a trifle precious, not wholly relevant to the urgencies of developing nations. If this explanation is even partially correct it remains an explanation, it is by no means a justification. For granted that it is possible, although questionable, to argue that an anxiety over air pollution is a luxury on which developing nations can ill afford to invest time and resources, it does not explain why problems associated with the control of such resources as soil, fisheries, aquifers, and forests and woodlands should have assumed so lowly a place in national plan documents. The point is that environmental resources are not restricted to amenities. The single unifying characteristic of environmental resources is their **natural regenerative property.** Environmental resources are resources which, unless they are severely depleted, are capable of being regenerated by natural processes. They are what are often called **renewable**

natural resources. Ironically, environmental resources are of critical importance to developing countries, more so perhaps than they are to modern industrial economics, for developing economies are for the most part still very much resource-based. One may therefore conclude that resource economics is not a luxury for these economies, it is a necessity.[1]

It is as well to articulate at the start the idea behind social cost-benefit analysis. Usually its role is taken to be restricted to investment projects. This is unnecessarily limiting. Economic options are not limited to investment activities. Tax policies, trade policies, food subsidies, relocations of populations, regulations on individual activities, are all options that governments deliberate over and choose from. Social cost-benefit analysis is merely a formal and consistent method of evaluating the implications of alternative policy options in the light of the variety of social objectives (or goals) that are judged to be worthy of pursuit. Thus growth in per capita income, increased employment, a more equitable distribution of income over assets, lowering the extent of absolute poverty, and so on. are all social objectives; at times consonant with one another, at other times in clash with one another. Economic option A might give a boost to a few of these objectives and a kick to the others, the reverse with option B. To judge between A and B is to rank them and the way to do this is to assess **how** A and B affect these multiple objectives. This is not an easy task, but it can be done, even though only approximately. Social cost-benefit analysis is an approach to **quantifying** the impacts of policy options and to reducing them to comparable figures, so that the options can be ranked. This has been done on many occasions for general investment projects in developing countries. It has rarely been done for environmental projects; that is, projects in which environmental resources are a major component.

In this article it is our purpose to present some of the salient features of social cost-benefit analysis and then to show how to conduct it systematically for environmental projects. Towards this we will first outline the logic of social cost-benefit analysis in a non-technical manner, highlighting the role played in such analyses by considerations of the distribution of income among contemporaneous people, as well as people separated by time. The next sections will apply these ideas to environmental problems. We will first identify certain pervasive forces that are known to trigger environmental degradation. Our illustrations will be based exclusively on **land resources**, and by environmental degradation we will thus mean loss in **land productivity**, or **land fertility**. Furthermore, our examples will be motivated by land degradation in Sahelian Africa. Having identified certain key forces that result in land degradation we will then suggest two classes of policies that could be used in confronting these problems. One of them has been much discussed in the literature and is often pursued in practice. This is direct investment aimed at increasing the resource base; that is, expending resources to regenerate, to revive, the productivity of degraded soil. Replanting trees and reseeding the soil are examples of such investment in land reclamation. We will, by way of a case study, discuss such policies.

The second type of policies consists of tax-subsidy schemes which are designed to alter the relative prices of commodities in such a way as to remove the pressure from land. The point here is that land provides a multiplicity of services. Among other things it provides crops, grazing resources and fuelwood. There are thus multiple "uses" of land. Acute problems can and do arise if an excessive reliance on one of these services degrades the capbility of land to provide the remaining services. For example, excessive harvesting of fuelwood can lead to the encroachment of deserts, thus making the land less capable of providing crops and grass. We should emphasize that considerations of the distribution of well-being will be very much in the forefront of our discussion and we will take note of the fact that people who rely most heavily on the resources provided by land (e.g., fuelwood, grass and crops) are the poorest in Sahelian societies, not the richest. Nevertheless, we will conclude that there are commodity-subsidy schemes which can meet the problems to a certain extent.

We will apply the methods of social cost-benefit analysis to illustrate how direct investment towards increasing the productivity of land can be evaluated. Our illustration there will be based on an actual report of a direct investment in improving grazing resources around permanent water supplies in Sudan. The report in question, UNSO (1982), is an impressive document. It provides financial estimates and indicates the kinds of benefits that would ensue, but it provides no social cost-benefit analysis. So we do not know whether on balance it was a desirable project.[2] For this reason we will present illustrative (i.e., fictitious) figures for the various features of a project and then show in a systematic manner **how** social cost-benefit analysis can be conducted. The idea is not to assess the actual project; rather, it is to use the project report as a **vehicle** for illustrating the framework of social cost-benefit analysis.

Social Cost Benefit Analysis - A Survey

(1) Introduction

In the previous section it was made clear that social cost-benefit analysis is a way of organizing information on the consequences of any societal change in order to evaluate the social desirability of the change. These consequences can in general be of very great variety including different real costs (i.e., the use of real resources), different benefits accruing to different individuals, etc. It is obvious that any useful information must be highly aggregated. The theory behind social cost-benefit analysis gives certain guidelines on how to aggregate the different consequences in a rigorous way. The purpose of this section is to give a very brief survey of these guidelines. For a more penetrating discussion the reader is referred to UNIDO 1972, Dasgupta 1983, Helmers 1979, Hufschmidt et al. 1983, and Mäler and Wyzga 1976.

(2) The choice of a numeraire

In aggregating different consequences we need a common numeraire by which we can measure all the various effects. Social cost benefit is developed around individuals and their valuation and it is therefore natural to adopt as a numeraire the equivalent income change. When applying CB analysis to project evaluation we are therefore interested in how each individual values the different consequences of the change and it is those values that should be used in aggregating all the effects. More precisely we will define these individual values in terms of a numeraire, consumption in general or income. Thus the value to an individual of having more bread would be what income he would like to give up to have more bread.

Our general definition is as follows: "The value to an individual of a consequence from a project is the amount he is willing to pay for having that consequence in case of a benefit or the amount he is willing to pay to avoid that consequence in case of a cost".

This definition is not satisfactory from a purely theoretical point of view because it is asymmetric between costs and benefits. For our purpose the definition will suffice, however, because any problem due to this deficiency can be handled by ad hoc methods.

Let us apply this definition to the valuation of increase in agriculture production in order to get the feeling of the use of the definition. Assume that there are perfect markets which are in equilibrium for agriculture products so that no individual consumer can affect the market prices. Then the value of increased agriculture production will be measured by the equilibrium prices, simply because the willingness to pay for different agricultural products will be equal to the price. If the willingness to pay would exceed the price, the consumer would find it desirable to increase his consumption until his willingness no longer exceeded the price and vice versa.

Thus we have reached the important conclusion that **whenever consequences from a project consist of goods and services bought and sold by consumers at certain prices, these prices represent the values that should be used in aggregating those consequences.**

However, this conclusion is based on the assumption that the individual can buy any quantity he demands. If he is rationed, either by disequilibrium in markets (unemployment f.e.) or by government regulations (foreign exchange controls f.e.) the willingness to pay may not be equal to the current market price and **the market prices must be corrected for the disequilibrium.** In particular, the trade offs between different time periods may not be characterized by accurate market prices, i.e., interest rates. Therefore a special discussion of the choice of a discount rate is needed. Moreover, if the demand for certain goods for the project is so great that the prices are affected, other corrections may be needed.

One particular price is of utmost importance in any CB analysis, namely the interest rate. The interest rate (or rather the discount factor) tells us the value of one unit of consumption one year ahead in terms of present consumption. There are many reasons why not even a perfect capital market may yield an interest rate useful in CB analysis. Therefore a special discussion is warranted.

Moreover, in view of the pervasiveness of high inflation rates and high dependence on international financial markets in most developing countries, it seems motivated to include a discussion of some problems in making real profitability calculations of investment in contrast to nominal calculations.

One particular feature is uncertainty, which creates certain problems in choosing a discount rate, and a few comments and some references will be given on this particular issue.

The most important reason for cost benefit analysis of environmental problems is the fact that most environmental consequences from a project are not valued at all on a market. Therefore special methods have to be developed in order to mimic the market when such consequences are to be measured, that is, to estimate the willingness to pay for environmental improvements or for avoidance of environmental degradation.

Finally, special consideration must be given to income distributional questions. In general the consequences of a change will affect different individuals differently and the social evaluation of the change may very well depend on who is benefitting most and who is bearing the costs. Thus it is very important to find out the distributional effects and to discuss in a rigorous way how to take these into account. In a later section we will give an example on how that can be done.

(3) Technical constraints

If the project is described in terms of its consequences as was discussed in the previous section and if the problem is to reject or accept the project, then there is no need to discuss technical constraints. They are already incorporated in the description of the consequences. Even if there is more than one project and the problem is to decide which of the potential projects should be undertaken, there is no need to worry about technical constraints as long as the projects are described in terms of all their consequences. This is the case for the example we will discuss later on.

However, when the number of alternatives is large it may not be possible to describe each of the alternatives in terms of their consequences in this way. In such cases it may be advantageous to represent the technological, biological and geological constraints in summary forms. The concept of a **production function** is particularly important. A production function describes the technical (or biological or geological) relations that must hold between inputs and outputs. In conventional economic analysis, a production function describes the relation between the quantity produced and the different quantities of factors of production utilized. We will assume that the reader is acquainted with this concept. The concept of production function has, however, a more general interpretation. In some studies, the production function will take the form of a **growth function.** A growth function describes the growth of a renewable resource as a function of the size of the resource, the harvesting of the resource and other factors that may be relevant. Growth functions are used in for example the studies on fisheries and ground water. Moreover, similar functions are used to describe the relations between the discharge of various pollutants in a water course and the ensuing water quality. However, in relation to this larger application, the term **environmental model** is more often used. The term environmental model also covers the diffusion models used by meteorologists to predict the ambient concentrations in the atmosphere at various points in a region from known atmospheric emissions of different pollutants. For a general discussion of production functions, and in particular growth functions, the reader is referred to Dasgupta (1983).

In some instances the conventional concept of a production function is not general enough. For example, whenever the interdependence between different economic sectors is judged important in connection with measures controlling environmental resources, **input output techniques** may be useful. Input output analysis is a technique which can be used to model the interactions between the different sectors in an economy. Basically it consists of the assumption that there are fixed proportions between the output in one sector and the various inputs (produced in other sectors) and the primary factors of production. The use of input output analysis in environmental economics is illustrated in Ahmad et al. (1984), in which the interested reader can find further references.

A development of input output analysis is **activity analysis.** This is an approach which is quite complex and we will not try to explain it here. However, it si based on a set of linearity assumptions and is also often called **linear programming.** Examples of the use of linear programming are given in Ahmad et al. (198).

(4) The use of market prices

One classical theorem in welfare economics says roughly the following: "Assume we have an efficient allocation of resources in the economy. By efficient we mean that it is impossible to reallocate the resources in such a way that no one is worse off and at least one individual is better off. If the preferences of the consumers are convex, if the production possibilities of producers are convex, and if all goods are sold and bought on markets at which no single individual can affect the prices, then the efficient allocation can be obtained as a market equilibrium by suitable redistributions of the initial wealth". Conversely, "any allocation that is the result of a competitive equilibrium is efficient given that all goods are sold and bought on a market".

These two theorems give the base for interpreting prices as measures of social opportunity costs. If the assumptions required in the theorems are satisfied, there would be no need for social cost benefit analysis, because the prices would carry all relevant information on opportunity costs and consumers' willingness to pay for goods.

The raison d'être for social cost benefit analysis is, however, the fact that these assumptions are not realistic descriptions of existing economies. Most of our discussions in this section will be concerned with situations where some of these assumptions are not accurate descriptions of reality. Such situations are denoted as **market failures.**

One of the assumptions necessary to make the fundamental theorem of welfare economics true is that **all markets equilibrate.** There are good reasons for some markets not to equilibrate, and this kind of market failure will be the focus of the discussion in this section.

Whenever a market failure is observed, there is an a priori argument for replacing the existing market price with a **shadow price** or **accounting price** which better represents the social opportunity cast of the good.

There are several treatises on calculations of shadow prices or the valuation of goods and factors of production for which equilibrium markets are lacking. One basic reference is the previously mentioned **Guidelines for Project Evaluation** (UNIDO 1972), and we will follow the methodology exposed in that monograph. F.L.C.H. Helmers in **Project Planning and Income Distribution** (1979) builds on the UNIDO Guidelines and gives simple expositions of some of the more complicated developments.

The basic idea is very simple, however, and we will try to develop the analysis for a very simple case, namely the determination of the shadow price on foreign exchange. There are many reasons (most of which are not valid) for a government to control the price on foreign exchange. In doing so the exchange rate will be different from that under price mechanism. The controlled exchange rate no longer indicates the true social value of a unit of foreign exchange and we therefore need a shadow price in order to handle transactions with foreign countries properly in a social cost benefit analysis.

What is, then, the social value of a unit of foreign exchange? That depends in general for what purpose that unit is used. One more unit of foreign exchange could be used to increase the imports marginally. The domestic value of the increase in imports made possible by an extra unit of foreign exchange is thus the social value of the shadow price on foreign exchange (assuming that the domestic value is calculated with "proper" prices). On the other hand, one extra unit of foreign exchange could also be used for reducing the exports and by that increasing the domestic supply of goods. In this case the shadow price is equal to domestic values of the increased domestic supply. If the extra unit of foreign supply is saved for future use, then the value would be the discounted future value calculated along the lines indicated above. In general, an extra unit of foreign exchange will be used both for increasing imports and decreasing exports, and the shadow price will be equal to a weighted average of export and import prices where the weights are based on information on how the extra unit is spent. For details, the reader is recommended to consult the above mentioned books.

Another important case is disequilibrium in the labour market. In general, the wage rate in the urban sector is higher than the subsistence level of income earned in agriculture in many developing countries. This creates an excess supply of labour which means that the ruling wage rate cannot convey proper information on the opportunity cost of labour. Instead a shadow wage has to be calculated.

Such a shadow wage should represent both the sacrifices made in terms of production foregone and the compensation for harder work and worse working environment (which may be negative if the project offers better environment and better working conditions than the alternatives). How this shadow wage is to be calculated is discussed in the specialized monographs mentioned above.

It should be noted, however, that the use of shadow wages (or any other shadow price) is really a second best solution to the allocation problem. If one uses a shadow wage which is lower than the prevailing market wage for the calculation of profitability for a particular project, one is not really comparing that project with all other projects but only with those that are using the same shadow wage. If all potential projects (including all private projects) were calculated with the same shadow wage, quite a different ranking of the project may result. It is therefore necessary to be quite careful in using shadow prices, trying to include as many as possible of the potential projects.

(5) Large projects

So far, we have assumed that the project does not change any of the ruling prices. That means that we could take the prices as given in evaluating the project. However, it is quite often the case that a project is so "big" in some of its dimensions that prices are affected.

The classical example is the construction of a bridge. Before the bridge is built, the cost of crossing the river is very high (in some cases infinite because it is impossible to cross the river). After the bridge has been constructed, the cost of one more crossing of the river is very low and can approximately be assumed to be zero. Here the price of crossing the reiver has changed drastically due to the project. Similarly, a dam for irrigation may reduce the cost of water supply, and the opening up of a coal mine in an arid region consuming huge quantities of water may increase the cost of water supply.

The problem we have to solve is to determine which price to use in project evaluations when the price changes as a consequence of the project. Intuitively, one feels that some average of the pre-project price and post-project price should be used. That this result is approximately true can be argued in the following way.

The demand curve for a good or service that is supplied by the project is shown in Diagram 1.

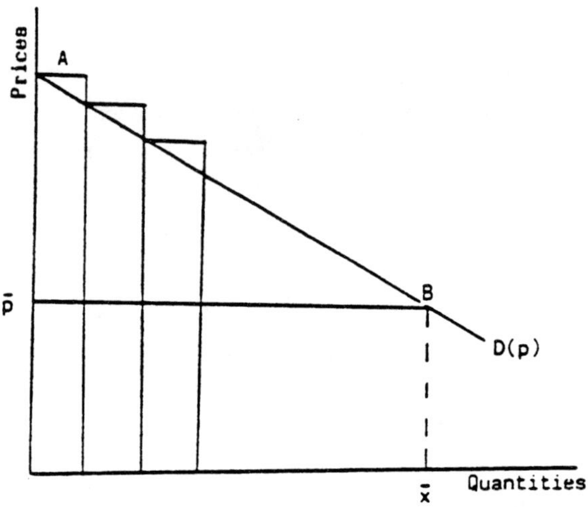

Diagram 1

Prior to the project nothing at all of the good is supplied, which can be represented by point A in the diagram. As a result of the project, the price will go down to $\bar{p}$ and the demand will increase from zero to $\bar{x}$. For the first unit of the good, consumers are willing to pay p_1. Thus the value of the first unit is p_1. They are only willing to pay p_2 for the second unit and their willingness to pay for two units is thus $p_1 + p_2$. Similarly, for three units they are willing to pay $p_1 + p_2 + p_3$.

Obviously we can continue this line of though until we have reached the quantity $\bar{x}$. But then the willingness to pay $p_1 + p_2 + \ldots$ can be approximated by the area under the demand curve, i.e., OABCO. This is the total willingness to pay for the quantity $\bar{x}$. If we assume that the demand curve is a straight line the area is equal to

$$\frac{(p_1 - \bar{p})\bar{x}}{2} + \bar{p}\,\bar{x} = \frac{p_1 + \bar{p}}{2}\,\bar{x}$$

i.e., the arithmetic average of p_1 and $\bar{p}$ times the quantity $\bar{x}$.

This argument is not theoretically correct, because price changes induce changes in ral income, which complicates the picture. However, Willig[3)] has shown that in general it is permitted to neglect these complications.

We have carried out the discussion in terms of a price change on a good that is supplied from the project. We could as well have looked at a price change for an input to the project and reached a similar conclusion.

(6) Evaluating environmental consequences

In the previous section we discussed cases where market prices do exist but do not carry correct information on social opportunity costs. There are, however, many projects that carry important consequences that are not priced at all in any market. Such consequences are generally known as **external effects**. Predominant among these are probably environmental effects. Pollution of the atmosphere or a water course leads to reduction of environmental quality such as hazards to human health, reduction of amenities, reduction in plant growth, decrease of the size of animal populations, etc. These effects on environmental quality have in general the property that they are not bought or sold on a market in common. There are very good reasons that we should not expect markets for such consequences. Environmental quality is in general a **public good** in the sense that an improvement in environmental quality for one individual means an identical improvement for other individuals. Thus, no individual has incentives to reveal a positive willingness to pay for environmental improvements, and voluntary agreements between the polluter and a victim of pollution would be highly unlikely. There are also other, more technical reasons for making such agreements unlikely. For a discussion of these, the reader is referred to Dasgupta and Heal (1979), and Mäler (1974 and 1985).

But even if there are no market prices on most environmental consequences from a project, it is still possible to apply the same basic idea in order to evaluate environmental consequences as the one underlying the functioning of markets. The value of an environmental improvement is thus defined as the willingness to pay on the part of the individuals to obtain this improvement.

As discussed earlier, we are by this definition sidestepping some theoretical problems concerning the difference between the so-called compensating variation and the equivalent variation. By the compensating variation we mean the amount by which we have to compensate the individual for a change in the environment. For an environmental improvement, the compensating variation coincides with the willingness-to-pay concept, but for an environmental degradation the two concepts are different. The equivalent variation is the amount which is regarded equivalent to the environmental change by the individual. For an environmental degradation the willingness-to-pay concept coincides with the equivalent variation.

Both these concepts are from a theoretical point of view correct measures of the individual's values. They are also in general not very different, but in some instances they may yield drastically different results.

Consider the case of an environmental degradation in the form of a dam construction which will flood a valley. One old man who has lived in the valley for all his life will be forced to move if the dam is constructed. His equivalent variation, or the amount by which one should reduce his income in order to expose him to an equivalent worsening of welfare, is bounded upwards by his income. On the other hand, his compensating variation is not bounded upwards and can be quite substantial. Although the two measures are radically different in this case, they both give valid measures of the welfare changes of the old man. The problem appears when we compare the welfare of this individual with the welfare of others, i.e., all those consumers of electricity who will gain by the construction of the dam. If we want to put special emphasis on the old man in the valley, we should choose to use the compensating variation, while if we want to treat him on an equal footing with the electricity consumer we should use the equivalent variation. For a detailed discussion on the concepts of equivalent and compensating variations, see Mäler (1985). In the rest of this section we will neglect the difference between the two concepts and continue to use the willingness-to-pay concept.

As there are no markets for environmental quality, it is not possible to measure the value of environmental quality directly from market data. Instead, more roundabout methods have to be used. Roughly speaking, these methods can be classified in four classes:

i) Whenever the environmental quality change affects marketable outputs (or inputs) that part can be measured by using market prices (market values).

ii) The change in environmental quality will induce shifts in individual behaviour. By measuring these shifts it may be possible to assess the willingness to pay.

   a) In the simplest case one can estimate the cost of a **substitute good.**

   b) Sometimes one can estimate the **opportunity cost.**

iii) Sometimes environmental quality changes will be capitalized in market prices such as property values and wages. By measuring these capitalization effects, the willingness to pay for an environmental improvement may be assessed.

iv) Asking the individuals exposed to a change in the environment for their willingness to pay.

These four groups are obviously no mutually independent but on the contrary to a large extent overlapping. For example, capitalization of environmental damage in land values obviously requires a change in individual behaviour, i.e., a desire to move from the degraded environment to a better one.

Before discussing these four groups of approaches, it is necessary to stress the importance of using a well defined concept of environmental change. These changes may be measured in several ways, depending on the specific problem at hand. Sometimes ambient concentrations of pollutants are used. Other times changes in biological populations or changes in landscpae are appropriate variables. One should also not forget that sometimes an environmental quality change is best measured as a change in a probability distribution. For example, air pollution giving rise to health effects is best attacked from the point of view of a change in the probability distribution of morbidity and mortality. This is illustrated in the case of air pollution epidemiology.

Quite often, an environmental degradation or improvement leads to a change in expenditures. For instance, increased particulate pollution may increase cleaning costs. Increased ambient concentrations of sulphur oxides may increase corrosion and thereby shorten the life of some materials. Increased acidification may result in reductions in the growth of biomass, both in agriculture and in forestry. All these examples are examples of **financial costs**, i.e., changes in expenditures that are associated with a change in environmental quality.

Generally, these financial costs are not difficult to estimate as long as we know the physical effects, i.e., the rate of corrosion, the reduction in biomass growth, etc. These are market prices that can be used to evaluate the cost of these effects. However, in general there are other kinds of costs simultaneously with financial costs. To a certain degree it is possible to reduce the noise from a highway by retrofitting the windows, increasing the insulation, etc. The costs of these measures are easily calculated.

However, this financial cost will not include the loss of amenity by not being able to open the windows, or by not being able to sit in the backyard enjoying a quiet evening.

Thus, there is often a need to complement the financial cost estimates with estimates of amenity losses. Methods of doing that have been developed and are presented in, for instance, Mäler and Wyzga 1976, Freeman 1979 and Hufschmidt et al. 1983. However, in connection with problems of soil erosion and land degradation in the third world, it seems reasonable to neglect these amenity losses. We will therefore not go into these methods any further.

Later on an example is given on the use of ii b) and the use of i), ii a) and b) is illustrated.

Finally, it must be underlined that some externalities are of such complex nature that it is still not possible to estimate in any reliable way the monetary values of them. It must be up to the decision maker to make the final judgement on the benefits from such externalities compared to other benefits and costs. In order to do that the externalities must, however, be described in a realistic way.

(7) Intertemporal trade-offs

By using the methods described in the previous sections, it is possible to reduce the description of the project to a stream of net benefits

$$B_o, B_1, \ldots, B_t, \ldots, B_T$$

over the different time periods (we disregard non-quantifiable externalities).

Some of these may be negative. In fact we will assume that B_o is negative, i.e., the society is experiencing a net cost in the first time period.

In order to make a judgement whether the project is worthwile undertaking, it is necessary to weight the net benefits in the different time periods with each other. This is accomplished by computing the **present value** of the stream

$$PV = B_o + \frac{B_1}{1 + r_1} + \frac{B_2}{(1+r_1)(1+r_2)} + \dots + \frac{B_T}{(1+r_1)(1+r_2) \dots (1+r_T)}$$

If PV is positive the project should be accepted, if negative it should be rejected. r_1, r_2, etc. are the interest rates by which future benefits are discounted to present values. r_1 is the interest rate in period 1, r in period 2 and so on.

$1/1+r_1$ can be considered to be the amount we would be willing to **give up now** in order to have one unit of the numeraire (say $) in period 1. In the same way $1/(1+r_1)(1+r_2)$ can be interpreted as the amount we are willing to give up now in order to get one dollar two periods ahead.

The discount factors can thus be interpreted as prices on claims for one dollar due to be paid in future periods. The present value of a stream of net benefits can then be interpreted in the following simple way. B_T is a claim on \$ B_T to be paid in period $_T$. Multiplying B_T with the price on a claim on \$ 1 in the same period gives the present value of B_T.

In general, there are no reasons to suppose that the interest rate will change over time (consistent intertemporal planning requires a constant interest rate, on this see Dasgupta and Heal (1979)). We will therefore assume that

$$r_1 = r_2 = r_3 \dots = r$$

and the formula for the present value simplifies to

$$PV = \sum_{t=0}^{T} \frac{B}{(1+r)^t} = B_o + \frac{B_1}{1+r} + \dots + \frac{B_T}{(1+r)^T}$$

The appropriate rate to be used in social cost-benefit analysis will be referred to as the **social rate of discount** or the **social interest rate.**

There are many controversies, both in the general discussion and among economists, about the choice of an interest rate. Some argue that some consequences should be discounted at a certain rate while others should be discounted at a different rate. In particular, it has been argued that environmental consequences should be discounted at a zero rate while other consequences should be discounted at a positive rate. The argument being that the environment belongs to future generations as well as the present generation and discounting the environment may deprive our children of their heritage.

In view of our interpretation of the present value, this argument is not valid. The comparison between the environment and other consequences in each time period has already been made when the net benefit B_t was computed. If the environment has a very great value for the generation living at time $_t$, that will be reflected in B_t. The net benefit should correspond to the trade off between environment and other goods that the generation living then is willing to make. This may be a very difficult thing to measure, but conceptually the net benefit should reflect the willingness to pay for the environment quality of the individuals living then. Obviously, we do not know the preferences of future generations but that is a problem we have to live with, quite irrespectively of the discounting procedures. We must, of course, use our best judgements about the willingness to pay for environmental improvements. But, and that is the important point, as soon as we have made an estimate of the value of the environment, all the consequences of the project are directly comparable, and all constitute claims on resources which should be valued at the appropriate prices, i.e., they should be discounted.

The biggest controversy, however, is about the choice of an interest rate. The literature on this topic is huge and there is no good summary available. Ultimately, the choice of a discount rate is a political one. That does not mean that economic factors are irrelevant in determining the social discount rate. On the contrary, economic factors are important and often decisive inputs in determining a social rate of discount.

Before discussing those factors, it may be a good idea to briefly discuss some reasons for a positive interest rate. One such reason is myopia, people are nearsighted and may lack the necessary imagination of fully appreciate future consumption. In terms of marginal rates of substitution between consumption in different time periods, it means that when consumption is the same in two periods, the rate of substitution is greater that one. There is fairly good empirical evidence that people in general are myopic, but that does not necessarily mean that society as a whole should use a positive interest rate to discount future net benefits. In fact, quite a few distinguished economists have argued that society should free itself from nearsightedness and discard this reason for a positive interest rate. On the other hand, if society is the set of its citizens it should reflect the preferences of these citizens. If the citizens are myopic then so should the society be. Obviously, this is a political question and an important political element behind the choice of the social rate of discount.

Another reason for having a positive interest rate is the productivity of capital and the growth of the economy. If the economy is growing, those who benefit in the future from a project will be wealthier and consequently value one more unit of the numeraire less than it is valued in the present. But this amounts to the same thing as discounting future dollars at a positive interest rate. Finally, the fact that investment in other activities in general are profitable means that using resources for a particular project has an opportunity cost that must be taken into account.

How all these factors should be combined so as to produce a social rate of discount is a very difficult problem and the reader is referred to specialized literature on that topic. Here it suffices to stress the importance of the opportunity cost concept. If we could neglect all other aspects of the problem of choosing an interest rate, then the opportunity cost would be the decisive factor. With perfect capital markets and no taxes or other interventions, the ruling market rate would be the appropriate one. However, with imperfect markets, corporate income taxes, etc., the situation is not so simple. One has to try to find out which other activities are displaced by this particular project and estimate the opportunity cost for each of these displaced activities. If consumption is reduced, the appropriate rate is the one consumers get on their savings. If private investments are displaced, it is the total (i.e., corporate taxes included) returns that matter. If both private investments and private consumption are displaced, one has to use some kind of weighted average of the two measures. The reader is referred to Lind (1982) for a more detailed discussion.

Before leaving intertemporal problems, there are three further remarks that must be made.

First, all the discussions so far have been in **real terms.** In real life, inflation is a pervasive phenomenon, which must be taken into account. In particular the nominal interest rates are substantially higher than the real interest rate. Although it seems self-evident, too much bad reasoning in the past makes it necessary to stress that it is **the real rate of interest** that matters, not the nominal rate observed on credit markets.

Second, although the present value criterion is the basic investment criterion, it is possible to use derived criteria. One such criterion, which in most real applications is equivalent to the present value criterion, is

the internal rate of return criterion. The internal rate of return is defined as the rate of discount that makes the present value equal to zero. It is thus defined as i in

$$0 = \sum_{t=0}^{T} \frac{B_t}{0 \quad (1+i)^t}$$

If i is greater than the social rate of discount, r, the project is worth undertaking, and if i is less than r the project should be rejected.

Third, very often a project is described by some average net benefit figure. To be more specific, assume that $B_o = -K < 0$ and that $B_1 > 0$, $B_2 > 0$, ... $B_T > 0$. Then it is often the case that annual depreciations D of the original investment K are calculated and the annual benefits are calculated as

$$B_1 - D,\ B_2 - D,\ \ldots,\ B_T - D.$$

If these are positive, the project is accepted and otherwise rejected.

This procedure depends crucially on the way the depreciation charges are calculated. **This technique should not be used** for project evaluation.

(8) Uncertainty

It is a truism that the future is concealed by uncertainty. The real problem is how to act in view of this uncertainty.

Let Y be the present value of the project. Y is now not known with certainty but is a stochastic variable with some probability distribution. It is thus possible that Y may be negative with a positive probability and positive also with a positive probability. Should the project be accepted? If we accept some very reasonable assumptions on rational decision making[5)] it is possible to show that there exists a **utility function U(Y) such that the correct criterion is to maximize the expected value of the utility function.** Unless we know the utility function (or at least some of its properties), this criterion is not very practicable.

However, in certain situations it is possible to conclude that the society should act as if it is **risk neutral.** By risk neutrality is meant that one is indifferent between a lottery ticket with a chance to win and a chance to lose and a certain outcome equal to the expected value of the lottery ticket. If society is risk neutral, the utility function is linear and we should try to maximize the expected value of Y. Situations in which this is true are when the present value Y is distributed among many individuals who share the benefits (if Y > 0) or the costs (if Y < 0) roughly equally[6].

If environmental changes are an important aspect of the project this result is not valid because of the public good nature of the environment. It is not possible to divide the environment up among the citizens in a way that the result requires.

Very often, uncertain projects are evaluated with uncertain variables replaced by their expected values, or **certainty equivalents**. This procedure, which is extremely simple, is based on some restrictive assumptions. One such assumption is that the technical constraints can be described by linear equations. As soon as the project results in irreversible changes in the environment, the use of certainty equivalents may lead astray.

In practical applications, the only way of including uncertainty may be to carry out sensibility analysis in order to find out critical values for the project design parameters. Anyhow, it is better to be explicit about the uncertainty than to compute expected values with a very low information content.

(9) Income distribution

The choice among projects has income distributional implications. Therefore these distributional implications are important in carrying out a social cost benefit analysis.

It is sometimes possible to include distributional values directly into the analysis by using different weights for different social groups. FOr a discussion of this approach the reader is referred to the book by Helmers (198). In a later section we will illustrate this method.

The most common way to handle income distributional problems is, however, to compute the impact of the project on different income groups. Then it is up to the decision maker to choose between the different alternatives, both with respect to their efficiency and their equity.

Soil Damage

(1) Introduction

Soil 'degradation', which is synonymous with loss in soil fertility (or productivity), is a generic term. It is convenient, therefore, to distinguish between, on the one hand, a reduction in soil nutrients and humus and, on the ohter, the erosion of soil (i.e., a loss of soil) by wind and water. To be sure, there are other sources of soil degradation, for example, salinization and waterlogging due to inadequate drainage. But in principle any such example can be fitted into our two broad categories.

Characteristics vary from place to place, but under normal circumstances soil productivity, in both these senses, is regenerated under natural processes. For example, Ehrlich et al. (1977) remark that: "The practical application fo the detailed study of soils is to discover both the potential and the limitations of different soil types for the production of vegetation, and the procedures appropriate to using each type as a permanent or renewable resource rather than as a "mine" to be exhausted and abandoned." (Ehrlich et al. (1977, p. 257). But in many locations the natural rate of regeneration is extremely slow and there may not be any means by which artificial augmentation can be made to occur. Thus, for example, salinization and leaching can be countered by creating adequate drainage and water retention facilities, but in the drylands a loss in top soil cannot be made good.[8)] Moreover, under natural conditions of vegetation cover it may take anything between 100 and 500 years for the formation of 1 cm. of top soil.

The causes of a fall in soil productivity are many, one class of cases being overgrazing (as in the fragile, semi-avid Sub-Sahelian West African rangelands); deforestation for firewood and wood products (as in Sahelian West Africa, northern Sub-Sahel and Sudan), and the clearance of steep slopes for cultivation. These typically lead to a loss in top soil and a decline in the capacity of the soil to retain moisture. Such are **direct** losses. There are, unfortunately, **indirect** losses in addition. Excessive evaporation occasioned by loss of soil cover implies a reduction in the rate of recharge of aquifers. For example, the sandy soils in the Sudanese rangelands have over the years eroded. As vegetation cover is sparse, sheep spill and gully erosion are quite extensive.[9)] Such erosion, together with the trampling of livestock can create a compact surface cover in the soil. Water cannot infiltrate easily and the runoff is high. This, of course, accelerates the process of soil erosion. It also affects the ground-water supply adversely.

As another example of indirect losses one may note that soil erosion in the uplands of watersheds can easily lead to sedimentation in the lowlands, reducing hydrological capability there and increasing the risks of floods.[10]

These examples highlight the effect of a loss of **soil cover.** A second major class of reasons for losses in soil fertility have to do with damaging agricultural practices. Blast and burn agriculture, excessive cultivation and imperfect irrigation practices often result in a loss of **soil nutrients** and in **progressive salinization.**

In the previous section we argued that environmental resources are **renewable natural resources.** Soil fertility is thus an environmental resource **par excellence.** If properly managed, it can be maintained almost indefinitely. If not, it can deteriorate, often to barren wasteland. At the practical level a matter of great importance is that soil degradation can occur through processes taking place at a distant geographical location (deforestation in the uplands affecting soil productivity in the lowlands). Such 'externalities' take time to monitor and to assess. But the **principles** of social cost-benefit analysis of projects aimed at recovering the productivity of soil remain the same, irrespective of the **source** of the degradation. In the previous section we sketched these principles and argued that environmental resources do not require of us to make any radical alterations to them. Thus, for example, the fact that such a resource is naturally regenerative if not overused will reveal itself in any good **feasibility** study of projects in which the resource occurs. One cannot modify the methodology of social cost-benefit analysis to take account of this fact. Just as engineering data must take into account the inevitable wear and tear of machinery, so must project feasibility reports incorporate data about the capacity of the soil to replenish itself over the annual cycle under the circumstances prevailing and those that are expected to prevail should the project in question be accepted. These are matters to be captured in feasibility reports of projects. Social cost-benefit analysis operates at a different, though related, level: the marshalling of the data presented in the feasibility report in a way which enables the decision maker to assess the impact of the project on social objectives.

Much work has already been done in exploring economic options open to the Sudano-Sahelian region, an area particularly suffering from loss in the capability of land to provide the essential services necessary fro their populations.[11] It is not our intention to review this literature, which is well known. We want in the following sub-sections to address a few of the issues that have not, to be best of our knowledge, been emphasized in the literature.

(2) Reciprocal externalities as a source of land degradation

There are several forces simultaneously at work and which precipitate soil degradation. These forces are in the main economic and social ones, unleashed by rapid population growth, the emergence of new political boundaries (thus, to take an example, prohibiting nomadic tribes from pursuing their customary migration patterns), civil wars that lead to major transfers of population and, for the purposes of this article, lack of effective markets for certain key commodities.[12)]

We have already noted above how soil degradation can occur via **undirectional externalities**, for example damages inflicted in the lowlands by deforestation in the uplands. Persons or groups engaged in deforestation are not required to negotiate with the damaged parties, who may be substantially far removed, geographically, from the uplands. These damages are not taken into account by the persons engaged in deforestation when choosing their courses of action. This alone would suggest a tendency towards excessive deforestation. The point we want to emphasize, however, is that in this example it is a lack of an effective system of possible agreements - or more technically, a lack of markets - which unleashes the process.

However, of possibly far greater importance in the Sudano-Sahelian region is the soil degradation resulting from **reciprocal externalities.**[13)] These have been much discussed in the literature, (see e.g. Dasgupta (1982)). A pristine form of this occurs in communal rangelands, where grazing rights are free to the herdsmen. There is an inevitable tendency in such situations towards excessive grazing. The fact that no individual herdsman has an incentive to restrict his cattle on a voluntary basis has been much noted in the literature. The rangeland, being a **common property resource**, invites excessive cattle, leading possibly to its gradual destruction and thus its desertification.[14)] The source of the problem here is the fact that the rangeland is common property and so no single person has any responsibility towards it. The rangeland is a free good when in fact its social value is positive. Each herdsman thus inflicts a damage on all the other herdsmen when he grazes his cattle - the remaining cattle have just that much less to eat - and it is a damage which he does not take into account when choosing how many cattle to graze and at what intensity. The externality is thus reciprocal.

A similar process is at work in the loss of soil cover due to the destruction of trees and shrubs for firewood needed for cooking and heating. This too is a phenomenon that has been much commented upon. Excessive deforestation here is due to firewood being a free good and the externality which is involved is, as with the case of overgrazing, reciprocal.

Traditional economic discussions of common property resources have emphasized the role that private property rights might play in ameliorating - or at worst stemming - the problem of excessive use. The idea here is to partition the (common) property into blocks and to award private rights to individuals for these blocks. The impracticability of such a proposal for shrubs and trees in the forest lands is quite obvious. So too is the idea of parcelling rangeland to individual members of nomadic tribes. In any event, in some parts of Africa nomadic tribes graze seasonally in land which is at other times cultivated by sedentary farmers. It is difficult to see the sense in which such land can be privatized, unless the herdsmen are found alternative locations.

There is an alternative proposal which has been much discussed in the literature: the imposition of user taxes on common property resources. The idea behind the proposal is plain. The resource, being common property, is currently free for use. Nevertheless, it has a positive social value, a positive shadow price. So if a price were charged for tis use, less of it would be used. If the price were set by the central authorities and levied by such authorities, it would be called a tax. If, in particular, the tax on each unit of the resource used is set equal to the resource's shadow price the resource would in principle be exploited at an efficient rate.

Notice both the similarity and difference between privatization and taxation as modes of control over the common property resource. If private property rights are awarded to the users they enjoy the surplus accruing from a more efficient use of the resource. If, on the other hand, the user-taxes were not accrued to the herdsmen they would not enjoy the surplus. Indeed, it can be shown (see Dasgupta and Heal, 1979, Chapter 3) that the herdsmen in this case would be worse-off than they would be under the common-property regime! With efficient user taxes the resource, to be sure, is efficiently exploited. But the distributional consequences can be horrific.

There are two major problems with user taxes as a control system: the first, as we have just noted, is ethical, the second practical. To take the first point one should note that users of common rangelands in Sahelian

Africa are often marginal farmers and nomadic tribesmen. They are among the poorest in society. Taxation for efficient use could pose a burden sufficient to impoverish them totally.

To take the second point, the **implementation** of a user tax is shot through with difficulties, for to be able to tax an activity it is a pre-requisite that the activity can be monitored. And monitoring costs may be very high when the resources in question are rangeland and timber. Both are spread over wide expanses, thus making it possible for illegal use. Exceptions apart, it is thus difficult to see how either privatization or taxation is a plausible control mode for common property resources of the kind that are involved in the process of land degradation and desertification.[15)]

(3) Direct investment and subsidy of substitute goods

Given that conventional suggestions regarding the social control of such resources as rangelands and timber in Sahelian Africa are not useful, what options remain? We want to argue now that there remain three broad categories of social control over the use of such resources. These modes are by no means mutually exclusive. In fact they can be used in parallel.

The first is the simplest and is often appealed to, especially by the users themselves, when they become cognizant of a common property problem. It consists of a cooperative move to agree to mutual restraint in the use of the resource. Over generations such restraints are often enshrined in social customs - for example the cessation of hunting during breeding seasons. Anthropologists have often remarked on the intricate network of rights and obligations that prevail within traditional societies dependent largely on a few natural resources. Nevertheless, one ought not to fall prey to misplaced romanticism. Changing population sizes and altered social moves cause strain on such customs and obligations, and the problem of common property resources can and does rear its head in a serious way. Regulations in the use of such resources - and by regulations we mean her **quantity** specifications -, even when hallowed by custom, may thus require reinforcement. Governments, through local officials, can play their part in such situations, advising, making information available, and adjudicating disputes. We would not wish to underplay the role of regulations on the use of resources. But regulations on their own are not enough. Firstly, regulations on their own are not necessarily consistent with **private incentives** and so, as with resource taxes, it may be exceedingly difficult to monitor infringements. In extreme cases local officials can be bribed.[16)]

The problem here is in the **implementation** of regulations. Regulations are difficult to enforce. Secondly, common property resource users are often the poorest in society. Today, in many parts of Sahelian Africa it requires eight to ten hours of work to collect firewood for a family's needs for a single day. It is difficult to envisage what regulations might mean in such situations. There is in addition a third problem with regulations, related to the second. Even if a complete ban on the use of the resource were imposed it may take a long while for the stock to regenerate. The users would thus be further impoverished for a long while in this case. This is a cruel dilemma. The users are the impoverished and cannot be asked merely to curtail their use, and the resource stock is dangerously low because of overexploitation in the past. We conclude then that the remedies which have traditionally been discussed by economists for the control of common property resources are of little use on their own in Sahelian Africa. We therefore turn to policy options that would be of value.

From the foregoing discussion two classes of policies suggest themselves for the purposes of increasing the stock - or **base** - of such resources as soil cover and shrubs and trees. The first is **direct investment** in the resource base, investment which requires expertise, and funds that the users themselves typically do not possess. In the next section we will discuss by way of an example the kinds of investment that would be involved and we will outline the method of social cost-benefit analysis that can be used in selecting from variants of such possible investment. The point to note for the moment is that direct investment is a way of rejuvenating a depleted resource base by injecting materials and creating physical circumstances that enable the potentially renewable resource to be a vigorous renewable resource.

But this, on its own, is unlikely to be sufficient, because the problem of the **commons** would remain; consequently the resource would remain vulnerable to excessive use. So a complementary policy suggests itself: the **subsidization** of commodities that are **substitutes** for the resource in question. The point here is this. By making substitutes cheaper individuals will have a natural incentive to rely more on the substitues than on the resource itself. Thus consider by way of an example the case of firewood. The wood itself is free, since the trees and shrubs are common property. But the private cost of **collecting** the wood is not nil, in fact it is extremely high today in Sahelian Africa. One reason, (there are other reasons), why in some parts of the world the rate of exhaustion of forest and wooded land increased rapidly during the decade of the seventies was the rise in the price of substitute sources of fire, such as kerosine oil. Subsidization of substitute fuel would reduce the private demand for firewood.

By way of a second example consider the destruction of common rangeland due to excessive cattle in the commons (see the next section). The rangeland being a common property, its use is free. But the cattle are not free and the price of cattle and the cost of maintaining them set limits to the number of herds cattlemen find privately worthwhile to own and to graze. If the private cost of cattle - relative to some other private asset - were to increase in value individuals would be prepared to own a portion of their assets in such other forms and not exlusively in cattle. The point we are making here is that the capital market in the rural areas of less developed countries is highly imperfect. In some cases nomadic people have no other form of capital to invest in, other than cattle, that is. Introducing alternative forms of assets which people can hold would allow them to diversify their portfolio, away from a concentration on cattle, a chief source of land degradation.

We conclude then that in addition to some limited forms of regulation two broad categories of policies need to be explored and pursued. The first consists in direct investment in the rejuvenation of the resource base, the second a subsidy on commodities which are a good substitute for those commodities that are involved in the process of destruction of the common property resources. This article, being concerned with the social cost-benefit analysis of investment projects, concentrates on the former. Towards this we use, for illustrative purposes, an interesting reforestation project in Sudan to see how the methods of social cost-benefit analysis can be applied to environment protection.

Replenishment of Grazing Resources around Permanent Water Supplies[17)]

(1) The project and its aims

In the northern parts of the regions of Kordofan and Darfur in Sudan the determining factor in the spatial distribution of people and their livestock is the availability of **permanent** water supply. In fact there are many locations where water is abundant in the rainy season but they are dry throughout the rest of the year, possessing no permanent source of water. Thus, it is only those locations capable of providing water over the entire year which become the loci of permanent human settlements, as well as centres to which livestock are brought from the surrounding areas on a regular basis for watering and grazing. This heavy demand on grazing resources in the vicinity of water points has led to their degradation. The carrying capacity of this land is thus low, and in order to be able to undertake grazing activities, cultivation and the collection of firewood, the population has been forced to locate itself at points farther and farther from the water supplies. Except for a few trees and certain kinds of shurbs, there is scarcely any vegetation left around many water supplies for use during the dry season, when it is most needed.

The town of El Odaya in Hammar District, in the region of Kordofan, is one such permanent water point which has suffered from severe degradation. In all, the population of 'greater' El Odaya is 40,000 people, 20,000 of whom live in the town itself and the remaining half in approximately 30 neighbouring villages. In El Odaya itself, the livestock population numbers approximately 2,000 cattle, 5,000 sheep and 7,000 goats. These numbers are further augmented by those which use the watering facilities in El Odaya on a regular basis, and includes the livestock of nomads which come to Hammar District each year to graze.

The total livestock population in Hammar District, at any one time, including those of nomads, ranges as high as 23,000 cattle, 200,000 sheep and 20,000 goats. The effect of these numbers of livestock is particularly damaging during the growing season when grasses are over-grazed to the extent that reseeding does not take place. The disappearance of the grasses, especially the perennials, further diminishes the capacity of the rangeland to support woody vegetation by reducing the relative humidity of the micro-climate and exposing tree seedlings to harsh conditions. In the case of fuelwood, inhabitants of El Odaya presently receive their wood supplies from sources up to 1.5 hours distance by truck. The existing farming systems of the area also contribute to desertification, as the soil is thus left all bare after the millet harvest and exposed to the rains and wind.

Very encouraging results at regeneration and sustained productivity of degraded areas such as this have been obtained in Sudan by protecting the areas around the water supplies during the rainy season, so that plants can build up their photosynthetic capacity, reseed, and thus provide forage during the dry season when it is most needed. It has therefore been proposed that **dry season reserves** around the town of El Odaya be established so as to provide improved grazing for a substantial proportion of the livestock population during the dry season. Recommendations concerning this proposal have been presented in a report titled "Management of Grazing Resources around Permanent Water Supplies", prepared by the United Nations Sudano-Sahelian Office (UNSO, 1982). In this section we propose to use this report as a basis for illustrating the use of social cost-benefit analysis of environmental projects.

The programme outlined in the report also envisages the creation of small watering points in selected villages in the surrounding area in order to alleviate pressure on land in the vicinity of the central wells. It is recommended that seven to ten new points be established according to the availability of accessible ground water and incorporated into a programme for management of the perimeters to be developed by the villagers themselves.

The project described in this document was formulated by a Government/UNSO/SIDA/FAO/UNDP mission to the project area in November 1981. It is envisaged that the project will provide for the development of aplan for perimeter establishment and management which could be replicated elsewhere in Sudan. Range rehabilitation is, however, a long-term process dependent to a large extent on rainfall, which is often erratic in such areas. Results during the life of the project may consequently not be entirely satisfactory from the point of view of the quality and quantity of vegetation produced if rainfall is low during the life of the project. The strategy developed, however, for planting and implementing range rehabilitation activities by the local population will be a significant contribution to desertification control which, if applied consistently over a number of years, will result in increased rangeland productivity.

The causes of land degradation in areas around permanent water sources are well known. The technocal solutions to the problem are straightforward: fencing off over-exploited areas allows them to regenerate with quite dramatic results. Moreover, this process can be accelerated and enhanced by reseeding and afforestation. Within a period of three years, land which had been virtually unproductive can be used for controlled grazing and, after some additional years, yield fuelwood and other possible tree products. This has been proven and demonstrated in such situations as the exclosure project near El Obeid and even in parts of a reserved area in the perimeter of El Odaya itself. The concept of perimeter fencing and control is over forty years old, and a perimeter project for El Odaya was begun some twenty years ago. Since then, however, this and other similar projects in the Western Sudan appear largely to have stagnated. Much of the fencing erected by government agents is no longer effective and the processes of grazing and erosion continue unabated. The initial step between the setting up of a technical solution and its acceptance and adoption by the local population has not been made.

The success of any development project, however sound its technical components, will depend on the extent to which the local people are involved in planning, implementation and evaluation. The technical solutions cannot be imported into the community as a ready-made project package; still less must they be implemented by external agents before they have beenfully understood and discussed by all the people concerned. This, the project report emphasizes.

Thus, the project envisages a major participatory role for the local population, in which problems and needs are analysed, technical solutions considered, a programme of rangeland management is drawn, implemented, controlled and monitored by the people themselves. In this way the project becomes a community, rather than a government, initiative: the people have a very real vested interest both in geting it started and in ensuring its continued and enhanced success long after (close) government or project staff involvement has ceased. This too, is something the report highlights.

(2) Limitations of the study

The objectives of the project are clear enough. They are, to quote the document:

(a) "a restoration of land productivity through improvement and systematic management of rangelands", and

(b) "contributing to the diversity and quality of agriculture and livestock production, thus promoting the area's overall economic and social development."

The document continues by suggesting that the "project will contribute to the attainment of these objectives by providing an economically viable self-help approach to rangeland management which, when applied on a broad scale, will significantly improve the quality of rangeland in the Sudan."

It is important to note at once that the precise organizational structure of the project - with its emphasis on **local participation** - is given as much weight in the document as increased land productivity. This emphasis on "social development", to use the document's phrase, is not something that is easily quantified, and it is of the utmost importance in social cost-benefit analysis to **not** try and quantify it. Such attempts would usually be disingenuous, justifiably unconvincing and thus ultimately self-defeating. Economists possess a catchword, "externalities", to capture such effects as improved organizational performance through training, group participation, and so forth. Having noted that there is a great del of evidence that local participation improves performance, and thus involves less expenditure of resources it is best to concentrate on those aspects of the project that can in principle be quantified, albeit approximately.

Here, the document is less than complete. It presents estimates of investment outlays and its breakdown between government contribution and UNSO contribution; but it does not attempt to quantify **any** of the benefits, nor indeed the flow of recurring costs. In particular, there are no estimates provided for the increased productivity of land. We do not in fact learn anything about the time flow of costs and benefits. This is one major shortcoming of the document, **when read as an evaluation of the proposed project**. The other weakness is this. Insofar as one can judge, the prices used to estimate investment costs are **market** prices, and these may well differ significantly from **accounting** (or **shadow**) prices in Sudan. Finally (and this is in fact subsumed under the previous point), no reference is made in the report to rates at which social benefits ought, in the judgement of the decision-makers, to be discounted. Choice of discount rates is, of course, rather critical for the evaluation of environmental projects, characterized as they often are by long gestation lags.

These foregoing remarks are not meant as criticism. We would not wish to suggest that the project is not worth pursuing. We are merely drawing attention to the fact that a proper economic evaluation of an environmental investment project requires very much the kinds of estimations and calculations that are required for usual investment projects. It is in fact entirely feasible to conduct social cost-benefit analyses of such projects.[18)]

(3) Project phases

We have noted that the project has two major components: (i) the establishment of dry season reserves around the town of El Odaya so as to provide improved grazing for a substantial proportion of the livestock population during the dry season, and (ii) the creation of small watering points in selected villages in the surrounding area in order to alleviate pressure on land in the vicinity of the central wells.

The project is seen as comprising four phases. First, a period concerned with the recruitment, training and establishment of the project staff in El Odaya. Second, a phase during which local development committees are formed for El Odaya and for selected neighbouring villages and the preparation of these committees of rangeland management plans. Third, the opening stage of the implementation of the rangeland management plans; and this includes construction (and repair) of fences and firelines; reseeding, afforestation and erosion control within the perimeter blocks as set forth in the management plans, and the digging of village wells. And fourth, the operative stage of the project, where previous investment capital is repaired and maintained, access to the perimeter blocks and use of grazing facilities and fuelwood is controlled. In short, phases three and four are what typically appear in feasibility studies of investment projects; namely, the investment period and the operating period. Phases 1 and 2 occur during the first year of the project, phase 3 during years 2-4 and phase 4 from year 5 onwards.

While the report recognizes that nomads are a significant proportion of the range users in the El Odaya area, the project places the environmental control on the sedentary population.

(4) Project investment costs and mode of finance[19)]

The project envisages three sources of finance. The investment costs are to be shared between the Government of Sudan and the United Nations Sudano-Sahelian Office (UNSO). Contribution of the UNSO is in the form of tied aid; that is it is tied to the project in question, and no repayment is demanded.

The project is expected to begin "operating" in about three years from the date of its initiation. In three years it is envisaged that reseeded areas will be ready for controlled grazing. In some eight years controlled fuelwood collection can also begin as trees planted in the initial year grow to desired sizes. The operating costs of the project are to be borne by the sedentary population. Thus, the project envisaged a self-financing scheme based on charges made by the management committee of the programme on livestock owners benefitting from the improved rangeland pasture. This tax revenue will be used for maintenance and expansion of the enclosures.

Notice therefore that the project in question combines at once both direct public investment in improving the productivity of land and taxation and regulation for the purposes of controlling resource-use once the investment bears fruit. (See the previous section.) This is what makes the project interesting for the purposes of illustrating the methods of social cost-benefit analysis of environmental projects.

In Table 1 below we present the time flow of expenditures expressed in US $ incurred by the project. Year 0, the initial year, combines phases 1 and 2. Expenditure in this year consists of salaries, construction of housing for project staff, purchase of equipment, and so forth. Total figure for this year is approximately $ 140,000. Direct investment in increasing the productivity of land and in the drilling of water holes occur during years 1-3. This is phase 3. Expenditures here include salaries, seeding, fending, drilling, and so forth. For simplicity, these expenses have been assumed to be incurred in a uniform manner over these three years. The annual expenditure is approximately $ 90,000.

Table 1. Expenditure incurred by project (annually), in US $

Year	0	1 - 3	4 -
Expenditure	140,000	90,000	50,000

It was noted above that the investment outlay is to be shared by the Government of Sudan and UNSO. We assume for the purposes of exposition that the outlay is shared **equally** between the two sources, and we note as well that the UNSO contribution is in the form of **tied aid**. Thus, it is a gift from an agency outside Sudan. The Government's contribution therefore consists of a flow of resources amounting to $ 70,000 in year 0 and $ 45,000 per year during years 1-3.

(5) Project benefits and costs

The project begins yielding benefits from year 4. The operating cost of the project (repairs, maintenance and salaries) is % 50,000 per year assumed constant over the future. These expenditure figures are summarized in Table 1.

We now turn to benefits. As noted earlier, phase 4 begins in year 4 (on the fifth year). The **net benefits** (to the sedentary and nomadic communities) from improvement in grazing and arable land are judged to be about $ 150,000 per year during the five years 4-8. These are the gains to be had from an **increase** in the productivity of land occasioned by reseeding, controlled access to pastures and so forth. The operating costs are $ 50,000 annually. These costs, as mentioned in the project report, are financed by the user-community through charges levied on the beneficiaries (for recall that the project is expected to be managed on a self-financing basis).

From year 9 (the tenth year) there is expected to be a market increase in benefits, sicne by then the trees which were planted in phase 2 will have grown sufficiently to allow for controlled exploitation as fuelwood. These benefits may well be the hardest to estimate, but they are not impossible to estimate. The savings enjoyed from having access to fuelwood in the proximity is the time saved in collecting it. Frequently at least one person per family is fully engaged in the collection of fuelwood. The savings achieved by growth in trees is the time and energy saved. An extreme limit of the value of these savings are 'market' wages for unskilled labour of the relevant category (e.g., female, child). At the other extreme is not to value the savings at all, which is absurd, since a **raison d'être** of the project is to make available cheaper fuelwood (in terms of time taken for collecting it) to the population in question. For the purposes of illustration it is taken that the benefits accrued from year 9 onwards are $ 350,000 annually. Of course, the operating costs remain approximately the same at the figure of $ 50,000 per year.

(6) Social benefits and costs

There are, broadly speaking, **three** categories of agents involved in the project: the Government of Sudan, UNSO and the population within the specified neighbourhood of the town of El Odaya (inclusive of the nomadic population affected by the project). Strictly speaking, there are other categories of persons affected, for example engineers and agricultural experts who are brought in from outside this area, but we ignore them, since they form a somewhat minor item in costs and benefits. Alternatively, one may wish to suppose that salaries paid to such categories of people are roughly equal to their opportunity costs and so, as a category of persons, they are unaffected by the project.

The partitioning of various persons and agencies into three categories is as much motivated by practical considerations as analytical urgency. In the previous section we noted that if income distributional considerations are to be taken seriously in project evaluation we must ask **who** loses and gains from the undertaking of the project. The poor people in the region in question are the patent targets of the project. But not all people in the region are poor and they too gain. If it is possible to obtain data about the benefits accruing to the poor and the non-poor in the region we would partition this population as well into at least two income categories. Usually, though, it is not possible to obtain such data. It is then inevitable that the evaluator lumps all persons into a single category - the "representative" person of El Odaya and its vicinity - and compares a dollar's benefit for such a representative person with a dollar's benefit for the government.

We are now almost ready to conduct our social cost-benefit analysis. But not quite. During years 1-3, (phase 3), areas of grazing land is, so the project report states, to be cordoned off from use. This inflicts a loss to the population currently using the land, for they will be foregoing benefits they currently enjoy. For the purpose of illustration we assume that this loss is worth $ 20,000 per year during phase 3 (years 1-3).

In Table 2 we present the flow of benefits and losses accruing to the Government of Sudan and the people of the region surrounding the town of El Odaya. We wish to evaluate the project from the point of view of the national objectives of Sudan. From Sudan's point of view the outlay incurred by UNSO is a gift. It is tied to the project. Hence its opportunity cost is nil. It follows that the UNSO contribution to the investment costs of the project should not be counted in the exercise. We are thus left with two categories of persons and groups affected: the Government and the people in El Odaya and its neighbourhood.

In what follows we regard income in the hands of the Government as the numeraire commodity at each date. Clearly it will be improper to value income accruing to the people in El Odaya on par with government income. It is a poor community we are considering and the project report explicitly mentions this poverty. Thus, in what follows we impute a weight of K, to every dollar of benefits accruing to the people of El Odaya. In other words, $ K in the hands of the Government is judged to be socially just as valuable as $ 1 in the hands of the representative person in El Odaya. K is an income distributional weight (see the previous section). Since it is more valuable at the margin to have benefits accruing to the representative person in El Odaya, we take it that K is greater than unity. For specificity, we assume that K=2. In general, since K is a **social weight** (see UNIDO (1972) and the first section above) and so it will be improper to fix on a value. As we argued above, political leaderships are unlikely to offer specific values for such social weights, or **national parameters**. As we argued earlier, it is best to conduct sensitivity analyses on projects by evaluating them on the basis of a wide variety of values for these national parameters. While we will use K=2 as a "standard" weight, we too will conduct sensitivity analyses.

Table 2. Flow of benefits and costs of project in US $ (annually)

Year		0	1 - 3	4 - 8	9 - ∞
Expenditure (materials, salaries etc.)	Government of Sudan	70,000	45,000		
	People of El Odaya and its neighbourhood	-	-	50,000	50,000
Benefits	People of El Odaya and its neighbourhood	-	-	150,000	350,000
Foregone Benefits	People of El Odaya and its neighbourhood	-	20,000	-	

The extent to which one ought to offer refined estimates of the social value of projects depends greatly on the quality of data at hand. It is unwise to attempt to be highly refined in one aspect of an evaluation when other aspects are crudely documented. Our aim here is to show how much more one can say about environmental projects with the crudest data than is usual. So we will improve upon the discussion in the project report by evaluating explicitly the time flow of social costs and benefits, having corrected for income distributional considerations.

The income distribution weight, K, is one national parameter. Another is the **social rate of discount**, discussed above. Much has been written on the subject of appropriate discount rates (see e.g., Lind (1982)). This too is unlikely to be given by the political leadership. Typically, one will wish to evaluate a project with different values assumed for it.

Let x denote the sum of the present values of the flow of net social benefits due to the project. Let r be the social rate of discount. On using Table 2 we now note that

$$x = -140{,}000 - \sum_{t=1}^{3} \frac{(45{,}000 + 20{,}000\ K)}{(1+r)^t} + \sum_{t=4}^{8} \frac{100{,}000\ K}{(1+r)^t} + \sum_{t=9}^{\infty} \frac{300{,}000\ K}{(1+r)^t}.$$

If x is positive the project is acceptable, if negative it ought to be rejected (first section). It is now easy to check that:

If r = 0,1 and K = 2, x = $ 2.958 million.

Thus with an income distributional weight of 2 and a social rate of discount of 10% per year the project is socially beneficial. This is useful to note because 10% is a high discount rate to use, and as Table 2 makes clear, the benefits occur only from year 4.

A more reasonable value of r is 0.05 per year. It is now easy to check that:

If r = 0.05 and K = 2, x = $ 8.499 million.

The project is highly desirable.

It is useful in such exercises to compute switching values of national parameters, that is values at which the project is **just** desirable. (See UNIDO (1972).) A great deal has been written on social discount rates, and less on income distributional weights. For this reason, at the intuitive level the effect of K on the social profitability of the project is not very clear. In what follows we let r = 0.05 and find that value of K for which x = 0. It is then easy to check that:

If r = 0.05, then x = 0 for K = 0.06.

This is a useful result to have. At 5% discount rate the project will look bad **only if** benefits accruing to the people of El Odaya are awarded a negligible weight (less than 0.06) as compared to benefits accruing to the government of Sudan. Specifically we note that so long as the weight awarded to the people of El Odaya is in excess of 6% of the weight given to government income the project, at 5% discount rate, is worth undertaking.

Notes

1. The link that can be drawn between resource economics and economic planning in developing countries is discussed at a technical level in Dasgupta, 1982.

2. We are by no means suggesting that it was not, merely that the document does not provide an answer.

3. Willig, 1976.

4. Dasgupta, Heal, 1972.

5. See for example Luce, R.D., Raiffa, H., Games and Decisions, New York, 1977.

6. Arrow, K., Lind, R., Uncertainty and the Evaluation of Public Investment Decisions, American Economic Review, 60, 364-378.

7. Helmers, op.cit.

8. In river valleys, on the other hand, the alluvial top soil is augmented annually by silt brought by the river from the mountain slopes.

9. See, for example, UNSO, 1982.

10. Deforestation of tropical forests can on occasion affect local precipitation. Loss of habitat can have substantial adverse effects on genetic material.

11. See, for example, UNEP, 1984.

12. Low and especially fluctuating rainfall precipitates these problems, as has happened in recent years in the Sudano-Sahelian region.

13. Rapid population growth is of course an important additional strain on the ecological system.

14. The problem of the commons was popularized in a well known article by Hardin, 1968.

15. By way of contrast one can readily see taxation as a mode of social control when industrial firms are engaged in the emission of pollutants.

16. For a vivid account, see e.g., Thompson.

17. This section is based heavily on UNSO (1982) and in parts reproduces passages from it. We have altered the numerical figures presented in the report so as to indicate that the points we wish to make in this section are solely for the purpose of illustration and are not designed to be comments on the report itself. Our reason for drawing upon an actual report is to show that the economic options we discussed in earlier sections were realistic ones.

18. A good checklist for the social cost-benefit analyses of soil and water conservation projects is: 'An Introduction to Cost-Benefit Analysis of Soil and Water Conservation Projects', by Jan Bojö; Southern African

Development Coordination Conference, March 1986.

19. Here we depart completely from the project report by constructing totally fictitious figures for costs and benefit flows.

References

Ahmad, Y., Dasgupta, P., and Mäler, K.-G. (1984), Environmental Decision Making, vol. 2, Hodder and Stoughton.

Arrow, K., and Lind, R., Uncertainty and the Evaluation of Public Investment Decisions, American Economic Review, 60, 364-378.

Bojö, J. (1986), A Review of Cost-Benefit Studies of Soil and Water Conservation Projects, SADCC, Soil and Water Conservation and Land Utilization Programme, Report no. 3, Maseru, July.

Bojö, J. (1986), An Introduction to Cost-Benefit Analysis of Soil and Water Conservation Projects, SADCC Soil and Water Conservation and Land Utilization Programme, Report no. 6, Maseru, September.

Cost-Benefit Analysis of Soil and Water Conservation Projects, Report from a seminar held in Maseru, April 1986, SADCC Soil and Water Conservation and Land Utilization Programme, Report no. 4, Maseru, July 1986.

Dasgupta, P. (1983), The Control of Resources, Southampton.

Dasgupta, P., and Heal, G. (1979), Economic Theory and Exhaustible Resources, Cambridge.

Dasgupta, P., Marglin, S., and Sen, A. (1972), Guidelines for Project Evaluation, UNIDO, New York.

Freeman, A.M. III (1979), The Benefits of Environmental Improvement, Baltimore.

Helmers, R.L.C.H. (1979), Project Planning and Income Distribution, Martinus Nijhoff Publishing, Boston.

Hufschmidt, M., James, D., Meister, A., Bower, B., and Dixon, J. (1983), Environment, Natural Systems, and Development, An Economic Valuation Guide, London.

Lind, E. (ed.) (1982), Discounting for Time and Risk in Energy Policy, Baltimore.

Luce, R.D., and Raiffa, H. (1958), Games and Decisions, New York.

Mellander, A. (1986), Cost-Benefit Analysis and Soil & Water Conservation Projects, Stockholm School of Economics, December.

Mäler, K.-G. (1974), Environmental Economics - A Theoretical Inquiry, Baltimore.

Mäler, K.-G. (1985), Welfare Economics and the Environment, in: A.V. Kneese and J.L. Sweeney (eds.) Handbook of Natural Resources and Energy Economics, vol. 1., Amsterdam, pp. 3-60.

Mäler, K.-G. (1976), Economic Measurement of Environmental Damage, Paris.

Willig, R. (1976), Consumers' Surplus without Apology, American Economic Review, vol. 66, no. 4, 589-597.

Chapter 14

ENVIRONMENTAL POLICY IN RELATION TO TERRITORIAL DISTRIBUTION AND PRODUCTIVE ACTIVITIES

M. Bresso

Introduction

This chapter is based on the results of a study conducted under the auspices of one of the CNR's goal-oriented projects.[1] The purpose of the study was to outline the situation with regard to environmental policies in Italy by connecting them to various economic and territorial configurations, and to come up with guidelines along which these policies could possibly be reoriented. For this purpose, the study established a framework of the situation in terms of various environmental components on the basis of existing statistical information, demonstrated the principal examples of deterioration or risk and then analyzed the policies of the entire nation and of some sample regions in order to show up their merits and limitations and to propose new guidelines.

It this chapter I shall comment on the subject that was the focal point of the study: the connection between Italy's various patterns of economic development and the phenomena of environmental deterioration connected with them, as well as the policies that may make it possible to control this situation (Bresso et al., 1988).

Environmental Conditions

The causes of environmental deterioration which may be attributed to man's activities can be classified in two basic categories:

- causes that modify the quality of natural resources; this category includes all types of pollution (air, water, land, sea) caused by emissions into the atmosphere or water or by solid wastes
- causes that degrade or consume the environment and natural resources, particularly the land, since they reduce its possible functions and/or aesthetic qualities. This category includes urbanization, large infrastructural projects, deforestation, attacks on areas of particular

---

[1] CNR (National Council for Scientific Research), Goal-Oriented Project, Economia Italiana, subproject 4, Diffusione territoriale dello sviluppo (The Territorial Spread of Development), Giorgio Fuà coordinator.

natural value (rural and environmental), destruction of flora and fauna, quarrying and mining, particularly when all these activities are carried out without any attempt to minimize their impact on the environment.

The actual dimension of these problems and hence the extent of the required rectification has often been underestimated in Italy because of an understanding of the situation based on little and unreliable data on the conditions of environmental factors. Analysis is never based on a long-term series of studies; instead, it involves a kind of "misleading composition" due to the difficulty involved in collecting all the data actually related to the consumption and deterioration of resources.

(1) The basic resources: air, water and land

There are many measuring stations that examine the pollution of these three recources. Unfortunately, they are usually engaged in the performance of tasks established by the various laws. Thus, these stations analyze the composition of atmospheric and water emissions rather than the quality of the resource itself. Indeed, the work of the various surveying networks is based on check-up, and there is no coordination between the networks that would allow for regional or national interpretation of environmental conditions. The fragmentary data available appear to demonstrate a certain tendency towards improvement in the quality of the air in large urban centres. This is mainly due to a reduction in the sulphur content of fuels, but is may also be due to a decrease in the most highly polluting industrial activities located around these cities. Nevertheless, extremely high levels of pollution are observed in stations placed at the nerve centres of urban traffic. In addition, warning levels are often exceeded during the winter or under conditions of thermal inversion during the hot seasons.

The reduction in pollution in large cities is probably also due to the territorial spread of productive activity, and hence of emissions, which took place in the 1970s. In fact, the estimated aggregate of pollutants discharged into the atmosphere for the entire nation only show a reduction in terms of sulphur oxides, while there has been an increase in all others.

Data on **water** quality are even less systematic. In addition, they do not demonstrate any improvement in surface bodies of water (lakes and rivers) nor in subterranean or marine bodies. Each time data is collected, pollution levels turn out to be alarming and often unexpected. In terms of subterranean waters, dangerous phenomena of accumulation (especially of pollutants of agricultural origin) are in course, and they cannot be dispersed except over very long periods of time and on condition that effective forms of prevention are found.

The condition of surface fresh and salt water is also alarming because of the fact that it cannot be corrected using the methods that have been employed hitherto. In fact, the majority of the investment is concentrated in the area of water pollution, especially as far as Investment and Employment Funds (F.I.O.) are concerned. This inability to have success is partially the result of contingent causes (30 to 50% of the purifiers installed do not function according to plan; many large cities are still completely without them; the industries that follow legal requirements represent only a part of those in existence). But there are also structural causes: a significant amount of waste does not pass through sewage systems (e.g., pollutants of agricultural origin), water is only partially purified and if it ends up in rivers without being diluted the conditions of the body of water continue to be extremely poor.

The waterways are actually a good indication of the general state of the environment, since they are the recipient of all forms of pollution at one time or another. Thus, a continuously functioning system which can analyze the quality of surface, subterranean and salt water appears to be essential. The system should be connected at nerve centres in order to be able to keep all of our waterways under constant surveillance.

The treatment of **solid waste** also shows significant aspects of impotence. Policies for air and water treatment have produced large quantities of purification muds which are difficult to store. Production and consumption cycles result in approximately 50 million tons of waste a year (approximately 15 of which are of urban origin). Once this waste is collected it is still a source of pollution or hazard, no matter what form of disposal is chosen.

50% of Italian municipalities have only a simple (and hence unsupervised) rubbish dump, and the forms of recycling are limited to just a few areas and a few substances (paper, more recently glass). Few regions have disposal plans established by law, while almost nothing has occurred spontaneously in terms of the treatment of industrial wastes, which could provide considerable possibilities for recovery. According to the law, the treatment of industrial waste is at the expense of the manufacturer. Thus, lawmakers evidently have not considered the fact that since these processes have no market value, the manufacturer who uses the worst treatment earns the most. This situation, as well as the fact that local communities accept (and not happily) only those plants that are guaranteed by public operators, will probably require a rethinking of this subject in terms of public regulation.

In recent years, the problem of contamination of sites by clandestine dumping of industrial wastes, usually located within a short distance of industrial parks (but not close enough to permit recognition), has become an enormous problem. A map of unacceptable dumping of industrial wastes is, in a certain sense, a map of external costs thrown back onto the community and the future during the boom years.

We no longer have a map of the external costs produced in areas with recent widespread development, but since many of the activities of Italian industry produce polluting wastes, it would be a good idea to begin to be concerned about it. The fact that the wastes are not found simply means that they have not been identified, and not that they do not exist or that they have been disposed of. There are few authorized incinerators, and their activity is anything but guaranteed.

(2) Territorial and natural resources

In this area, the only data available concern land consumption. Over the course of thirty years, the number of residences has tripled; the number of roads and infrastructures has multiplied several times, and costs have probably increased by over 50% (Min. Amb., 1987).

During the same period, an agricultural surface area equal to three times the size of the region of Liguria has been consumed. The widespread deterioration produced by tourism, which consumes rare environmental resources, cannot be quantified, but nonetheless it has affected all resorts.

Productive, uncultivated land and urbanized surfaces represent 20% of the total surface area of the country, which is equal to the area covered with forests and coppices. The remaining 60% is covered with agricultural land which represents land "hunted down" for other uses without any protection. Agricultural uses are themselves sources of severe pollution.

More specific information, which is all the more disturbing since it involves urban areas, may be added to these general data. The prediction for the Milanese metropolitan area is that urbanized land will soon exceed 50% of the total (I.R.E.R., 1980). A study on 110,000 ha of surface area and 7.3 million inhabitants (It. Urb., 1980) has demonstrated that the extent of urbanized area has more than doubled during the thirty-year period covering 1951-81, with peaks of 300% in some metropolitan areas.

Standing up to these assaults, there is a modest 4% of protected area (parks and preserves), with a peak of 6-7% in Piedmont.

Trends for the future are disturbing: widely accepted development models that encourage widespread settlement and individual residences may have allowed our country to avoid overgrowth in large metropolitan areas (which could be defined as "black" zones on our territory), but they have also involved a more extensive, though lighter, compromise (these areas could be called the "gray" zones representing deterioration of the land and the environment).

Despite the fact that the majority of municipalities have adopted an urban-development plan, and the regions have often produced documents relating to territorial planning, the guidelines contained in these documents are still functional distribution of activities over territory and not the protection or preservation of environmental and territorial resources.

Processes of deterioration, even when their origin varies, also appear in rural portions of the lowlands, which are increasingly characterized by monocultural practices that impoverish the land and result in its chemical or organic pollution brought about respectively by fertilizers, herbicides, pesticides and landless stock farms. Problems connected with the abandonment of cultivated land, poor maintenance of forests and consequent danger and degradation predominate in marginal areas.

In all the cases discussed above, we are not faced with acute situations, but rather with the slow deterioration of our country and with problems that extent throughout the nation for which the Galasso act (no. 431/85) attempted an inversion of the trend (especially culturally). The act concerns protection of the countryside, that is, of the complex stratification over the land of various life forms and cultivation and settlement practices, which is difficult to define but perceptible to all. The difficulties that are encountered when drawing up plans for rural areas are connected with the most complex of all the problems that our civilization must face, that of leaving a trace of itself which does not destroy evidence of the past and at the same time is able to create new aesthetic values and new forms of compatibility between human activities and natural or anthropic environments from previous cultures.

Desirable and Undesirable Trends in the Development of Settlements and Productive Processes

An evaluation of the trends described above, both in terms of densely populated zones (metropolitan areas and zones of widespread development) and less densely or even depopulated zones (rural, mountainous and hilly areas) has made it possible to demonstrate the problems that exist in terms of pollution or the use of land and environmental resources.

In brief, there is a need to control the processes of spread already in course, since they are responsible for environmental compromission which may not be immediately perceived but against which prevention is the only cure. An attempt has also been made to demonstrate which spontaneous trends seem to lean towards less environmental compromission and thus deserve to be the object of incentives, and which, on the other hand, must be controlled.

Some of the most interesting **desirable trends** are psychological, for instance the growing environmental demand on the part of the population, the demand for increased urban quality, and the increasing influence that environmental organizations have in decisions that affect the environment. Positive trends have emerged from important industrial sectors, which are lightening and reducing consumption of materials and energy. The increased importance of immaterial goods in terms of production and research developments in the biological sciences are also positive occurrences. Positive territorial trends also include the favourable aspects of the spread of productive activities over a greater area. When this is accompanied by modifications in productive processes in terms of a decrease in their impact on the environment, this could help ease the pressure on the most seriously compromised zones. Other territorial trends include the spatial and temporal diffusion of tourism, the push (also on the part of the E.E.C.) towards more appropriate technologies for agriculture.

In addition to these positive trends, other **undesirable trends** which increase the level of deterioration and compromission have been discovered. In terms of productive sectors, problems include the difficulty in controlling polluting emissions from small and widespread industrial activities, the absence of control over new products which are potentially dangerous (particularly in the chemical and genetic engineering sectors), the replacement of micropollutants with macropollutants and the spread of landless stock farms. Negative trends connected with territorial processes include the spread of individual dwellings, the resumption of massive infrastructural investments, the growing demand for high quality vacation homes and the renewed pressure of tourism on natural resources which have not yet been destroyed. All these phenomena lead to an increase demand for land in areas which are still well-preserved and, despite certain signals, are far from being the object of careful evaluation of the environmental impact.

The Status of Environmental Policy

The second phase of the study analyzed the environmental policies currently in force in order to bring to light orientations which could improve the situation.

In Italy, environmental policy has never been the subject of a specific planning document. The only mention in this regard is contained in some parts of "Project '80", which contains indications about depollution, defence of the land and preservation of our historical and artistic patrimony, in addition to an analysis of urban hierarchies and related proposals for rebalancing. The document's approach is naturally connected with the environmental ideas of the time, and thus the actions envisioned are mainly connected with purification.

Since that time no other planning documents have been approved (indeed, "Project '80" remained purely an analytical document).

The establishment of the Ministry of the Environment is a recent occurrence, and therefore there is no plan for the environment or specific documents related to the most important questions. Indeed, the "Report on the State of the Environment", which is required by the same law which established the Ministry, has only been presented in the form of an incomplete draft.

The Italian Regional Governments have behaved differently. Presidential Decree no. 616/77 gave them considerable power in terms of the environment. Known as planning bodies, they have made a considerable effort to place their actions within a programmatic framework. Practically all regions have approved development plans, and many regions have also approved territorial models. These documents always contain paragraphs devoted to environmental policies. Some regions, such as Lombardy, have also drawn up specific documents relating to the environment, and almost every region has plans for particular trouble spots (e.g., water and waste). The experiences of the various regions have been very different, but in general the main problem appears to be the inability of these plans to guide concrete legislative and administrative action. This may have been inevitable in a situation in which the absence of a national plan makes it impossible to proceed with the financing of regional policies on the basis of objectives and criteria originating from it. But the same thing has often happened in the relationship between the regions and local organizations. The objectives of these plans have been so unclear as to leave room for traditional decision-making criteria. The great discrepancy between plans and administrative practice has appeared as a result of the more innovative policies, which would require a greater ability to govern.

It would be a mistake, however, to dismiss the regions' planning experiences as all part of some "dream books". Some regions have executed significant portions of their plans, they have learned to correct the errors of the initial phases and have begun to understand their specific environmental and territorial situation. Therefore, the problem is not whether to give up planning activity, but rather to learn through subsequent adaptations to increase the efficiency and the ability to act retroactively between concrete experience and planning.

The difficulty in dialogue between the national and the regional governments has also been demonstrated by the trend in public spending on the environment. The only "inevitable" part appears to be that portion earmarked for "natural disasters", which are often external costs of past deterioration. The other items seem to be residuals because, as a result of the fairly constant level of allocations, spending on these items has an inverse relationship to that earmarked for disasters (Min. Environment-Pub. Econ.). Thus, the financing for various activities oscillates, because it is influenced by the various "emergencies" which occur.

Basically, there is an absence, on the national level, of a comprehensive programme that would establish which activities should take priority in which sectors and what the rate of overall reduction of pollutants should be and by using which methods and which allocations. Subsequently, these indications should be compared with regional environmental plans, which could be financed in their entirety and not item by item (although not all their requests could be met, of course), as long as they conformed with general objectives. Out of this would come a double obligation for the regions: first of all, to come up with plans that were feasible and well-coordinated with the general objectives, and secondly, ot actually carry them out and demonstrate their effectiveness in terms of their established goals; otherwise they would lose subsequent financing.

In summary, it may be said that the task of guidance and coordination of environmental affairs, which belongs to the national government, has mainly been carried out through specific legislation and the use of allocations aimed at certain policies, partially consisting of F.I.O. funds. Programmatic guidance and verification of the correspondence between objectives (implicit in the regulations and allocations) and concrete results, and between the latter and their effect on the environment, has been almost totally lacking.

On the regional level, it cannot be denied that a programmatic effort has existed, although it has been partially frustrated by the absence of dialogue with the central government. On the other hand, the existence of tools to verify the efficiency and efficacy of activities (both legislative and financial) has been lacking (cf., the reports on sample regions).

With regard to the problem at hand, the relationship between environmental policy and territorial specificity has suffered from the substantial absence of allocations, for specific local problems, of funds, which were distributed to the regions according to standard parameters. In the case of F.I.O. resources, there has been appropriation, but the cost/benefit analysis has been limited to microeconomic aspects and to the regions' overall needs.

Signals of an inversion of this trend have appeared in the form of the recent use of the "declaration of a high-risk environmental area", which was created by the law introduced by the Ministry of the Environment. This declaration has been used for some highly polluted or deteriorated zones, such as Naples, the Bormida Valley, the Lambro and Olona Rivers, and for the Po basin project. In these cases, the allocated funds are used for a specific area problem, although these activities may still be considered emergencies and not part of a plan that maps out territorial priorities. In addition, it is not clear why these special activities have to be supervised by the national government.

Methods for public action (at all levels) related to the environment have generally remained connected to the policies of the 1960s, that is, to activities begun after the fact: purification, depollution and dumping. Today, reclamation in those areas in which major problems have accumulated (in effect, "artificial" disasters) has been added to the list of activities. Little or nothing has been done to redirect environmental policies towards prevention and recovery/recycling, despite specific indications (sometimes even directives) on the part of the E.E.C. in this regard. The few attempts made in this area have included the law on energy conservation (with little financing however), the regulation on the reduction of sulphur content in fuels, others on phosphorus content and the biodegradability of detergents and still others on non-biodegradable packaging which should no longer be permitted after January 1, 1989. The use of methods such as emission taxes or standards (with the extremely modest exception in the case of the law on water) or incentives for clean technologies and/or disincentives for polluting technologies and products are completely non-existent. In fact, the financing of depollution as opposed to research into less polluting technologies has certainly not been a technology-forcing policy.

Even the introduction of preventative evaluation techniques to examine the environmental effects of large investment projects (or possibly general and specific plans as well) (E.I.S.) has been limited to an occasional study on the impact of large infrastructures, despite the existence of an E.E.C. directive to this effect.

The Discrepancy between Cultural Acquisitions and Concrete Environmental Policies

One of the acquisitions which has been consolidated in over twenty years of evolution in environmental policy in developed countries, is the fact that environmental protection as a segmented policy often encounters limits. Experience has shown that segmented policies of purification/ depollution too often produce a transfer of the pollutants from one environmental receptacle to another. According to the E.E.C., it is necessary to act "in order to prevent the onset of environmental problems deeply influencing industry, agriculture, energy production and transport". Therefore, environmental policy must become a part of politics and economics, as well as a sort of large filter through which all segmented policies should be shifted.

This approach is now accepted by all international organizations and by experts in the field. However, environmental policy, at least (but not just) in Italy, still tenaciously clings to the previous type of activity. It is important to attempt to understand the reasons for this:

- first of all, there is the relative simplicity (apparent) of depollution techniques; all that is necessary is the addition of a kind of "tail" to the production process, which gives rise to new sectors, among other things
- second, the power which the government and administrations have within a market system should also be considered; it is relatively easy to impose qualitative standards for dumping waste at appropriate sites (apart from then being able to make sure they are respected). It is much more difficult to try to orient productive processes and products. Indeed, this leads us into the complex area of competitive mechanisms, the forced orientation of technology, the prohibition of the use or production of certain goods. All of these are areas in which companies do not happily accept interference

- third, there are certain types of emissions and wastes which are apparently unmanageable a prior: sewage dumps, urban wastes, etc. It is possible to act on some aspects, but it is still necessary to dispose or purify the waste
- fourth, the "vertical" organization of public administration should also be considered. The Ministry or the Regional Departments of the Environment have quickly been identified as groups which are mainly responsible for purification activities and related regulations and controls, and funds are allocated to them for this purpose. If the pressure from manufacturers of antipollution equipment (who have found a smaller market than expected due to extremely widespread disregard of these regulations) is also considered, it is easy to understand why purification, depollution and dumping continue to dominate environmental policy. Recently, the reclamation of deteriorated sites has been added to this list, and it has unleashed proposals for action and the establishment of companies that "specialize" in activities whose scientific and technological aspects are not yet known and whose concrete experience is extremely flimsy.

Despite the discouraging information provided above, there is something positive to be said: the widespread conviction among technologists that industrial production has reached a turning point. It would seem that it is becoming lighter, reducing the waste of resources and energy and attempting to contain dangerous emissions. Could the "technological factor", which Barry Commoner identified in the 1960s as primarily responsible for the increase in pollution (with respect to population growth and per capita production), have started to demonstrate an inverse tendency? How much of this is due to spontaneous trends and how much to the strict regulatory policy of past years enacted by countries that are leaders in technological innovation? Does this also mean that countries such as Italy, after having imported polluting technologies, will also import "lighter" technologies?

All of these questions are of great importance for the establishment of an effective economic-environmental policy whose objective is to reconcile these two areas of public action, which are often divergent. A policy aimed at encouraging and guiding processes which were not counterproductive in terms of its objectives would certainly be easier to manage.

Elements for the Integration of Environmental Policies and Economic and Territorial Policies

This is an extremely vast subject; in this section I shall attempt to explain those environmental policies which seem to be most interesting in relation to the activities indicated above.

(1) Policies for environmental reclamation and pollution control

Currently, the majority of regulations in this area concern the quality of emissions (standards for dumping in water, air pollution around factories). This means that the overall quality of the bodies of water or the air is not dealt with. Furthermore, regulations concerning air, water and waste are not integrated, and this encourages the "transmission" of pollutants.

Therefore, the first objective is a comprehensive reduction in air and water pollution and a decrease in the production of waste (e.g., a given annual percentage). How much reduction is necessary? Reduction should continue until the residual flow is equal to the environmental receptacles' natural capacity for absorption. This involves another aspect: the reduction and gradual elimination of substances which cannot be "recognized" by the environment and thus cannot be absorbed. Policies in this regard are very diverse. This is because for the air it is necessary to define comprehensive emission limits since there is really only one receptacle, whereas for water and solid wastes it is necessary to determine the specific capacity of various areas (water basins, the characteristics of the land used for dumping).

In addition to the traditional strategy of acting after the fact, which cannot be eliminated in the short-run, there are also two other policies:

- recovery/recycling/conservation, such as the treatment of closed circuit refluxes and the recycling and recovery of waste from production processes; reduction in the quantity of urban waste through obligatory forms of collection; reduction of atmospheric emissions through the recovery of heat and energy conservation.

Currently, policies of economic support or fiscal measures are required to direct the market towards significant recovery processes. It is important to organize markets for the recovered materials, increase the cost of water to encourage it reuse, and establish regulations for products that make their recycling easier and less costly. It is also necessary to encourage research into recovery technologies, which are not always available or reliable.

- modification of production processes and of products, so that they pollute less or emit less dangerous pollutants. This policy is the most interesting, but also the most difficult to pursue. Possible methods are:

(1) technology-forcing, or the forced orientation of technology towards less polluting practices, attained through increased pressure on industry in order to encourage manufacturers to look for solutions which are alternatives to purification.

(2) the use of incentives and disincentives; for example, incentives for those who reduce their water and energy consumption or for research into products to replace plastics, deposits on glass and other packaging products to encourage recycling, incentives for less pollutant agricultural practices, etc.

(3) actual prohibition of the use of certain processes or products; for example, prohibition against the sale (or use or production) of certain chemical products used in agriculture, prohibition against the use of non-biodegradable plastics for packaging, etc. Application of this method would be limited to very high-risk cases when products have already been authorized, but it could become standard practice during the preliminary analysis phase of the environmental impact of new processes or products.

(2) Policies for the reclamation of damaged environments

This is the other side of the coin in terms of the damage inflicted on land and bodies of water during past years. Given the extremely serious level of pollution of some bodies of water, geological strata and vast areas, a policy of reclamation has been proposed with increasing insistence. It should be noted here that there is the same risk mentioned above regarding pressure from manufacturers of "recycling" techniques for purification who propose massive clean-up operations. The problem is real, and there is no doubt that reclamation is necessary. However, at the moment the scientific meaning of words like "reclamation" and "decontamination" appears to be unclear. The causes of deterioration may be many, and it is first necessary to discover them. Work must be done on catchment basins, but also on productive (pollution producting) and territorial (areas where waste is stored) levels as well. It is necessary to begin an integrated evaluation of the current flow of pollution in order to analyze its compatibility with the land, water (both superficial and subterrenean) and air, to decide whether to modify the processes or stop incompatible activity. Only later will it be possible to begin reclamation of deteriorated areas in the form of clearing or draining of inadequate waste and mud dumps, inspection of the status of toxic dumping in waterways and possible removal and examination of the status of geological strata, flora and fauna. Planning of required activities, which will not be short-term propositions nor always possible to carry out, is also a later step. In many cases, the only "reclamation" possible will be the reconstruction of a balance between human activity and the environment, while leaving the real healing work to nature. The risk of hasty, costly and unjustified operations is great, given the extreme fragility of the material and the extreme attractiveness of the immense funds available.

(3) Protection of natural and territorial resources

I have already mentioned how dangerous effects on territorial resources may result from both excessive pressure due to use and conditions of abandonment and depopulation. In the case of land, operations undertaken once deterioration has begun are not always possible, and they may be very expensive and difficult. Therefore, land overuse is irreversible, and this should encourage us to look for preventative forms of compatibility among environmental and rural protection (which are not the same things), the current functions of the land and the actions that are to be undertaken.

In general terms, territorial policies should no longer plan the distribution of activities on the basis of economic and functional criteria. Instead the protection and preservation of environmental and territorial resources as their orientation should be used. This would require useful territorial and environmental information systems which would make it possible to evaluate choices on the basis of criteria connected with the quality of resources, their current status and potential risks.

The tools available to a planner in order to institute effective protection of territorial resources include:

- territorial and urban planning
- rural planning
- creation of parks and preserves
- environmental impact assessment of plans and projects

This is a complex set of tools which must be used to operate within a unitary vision, on different territorial levels, without creating an overly complex and therefore ungovernable machine.

With the partial exception of park policy, which has allowed research into ecosystems and the teaching of environmental science when it has been enforced, territorial policies have basically remained theoretical models.

They have not managed to significantly influence the dynamics of land use. Urban planning, which has also spread widely, has not managed to add quality to cities or their buildings.

Rural planning has only recently been established, but it gives the impression of continuing to be a kind of census-taking of regional rural values, while the proposed plans appear to be incapable of forcefully directing the modifications in course on the land.

Under these conditions, too much hope is placed on methods such as environmental impact studies, which are by their very nature cognitive and hence preliminary methods to decision-making about the use of land and natural resources. The E.I.S. demonstrates the predictable modifications that will occur in an environment following the execution of a project; it does not help decide whether to execute or not. Its main positive effect is that it encourages improved design from the standpoint of compatibility with the environment (or planning, when the E.I.S. is concerned with planning tools), in addition to making it possible to confirm predictions. It also makes subsequent evaluations and related public decision-making easier. In the absence of simple but sure methods to guide territorial choices, the E.I.S. is given a sort of planning role a posteriori, which risks turning it into a complicated, costly and useless tool.

The battle over alternative uses of the land and of primary resources (air, water, etc.) may concern the use of water in the city as opposed to the countryside or the use of valleys for infrastructrual purposes, to name a few examples. The interests of city-dwellers in preserving nature in underpopulated areas are in contrast with those of area inhabitants who wish to promote development. No municipality is willing to accept a dump in the name of defending its own environment, but no one is willing to let us create a national park either. The conflicts between inhabitants of an area and the industries located there are numerous. It is clear that an E.I.S. may demonstrate the advantages and disadvantages of a choice, but only territorial planning is able to guide decision-making in the case of divergent interests.

Therefore, it is necessary to relaunch the idea of territorial planning by turning it into a flexible tool, a means for the interpretation of environmental compatibility on the territorial level.

(4) Operative levels and tools for action

Execution of the policies that have been outlined above, alternated with pure depollution, is a complex undertaking. Thus, it is necessary to determine the proper level and the correct tools for action. Policies favouring recovery, recycling, reclamation and conservation (together with territorial policies) should typically be managed on the regional level, although national laws will sometimes be necessary to orient behaviour and production. In addition, these policies are often already "written down" in many regional documents. How can they be made into more than just paper? One method could be that of Regional Agencies, public bodies with the possible participation of private interests, but with autonomous management. Indeed, it would be necessary to begin a series of studies and experiments, stimulate action and plan activities; this would require a high level planning group that could supply guarantees to all interested parties. These Agencies would operate in the areas of energy and water conservation and waste reduction, in tune with the above mentioned objectives, and they could construct significant models for action.

Policies regarding productive processes and their products are typically within the competence of the national government. It is necessary to determine which activities have a considerable impact on the environment, define the priorities for action and force industry to make a significant effort towards redesign with environmental effects in mind by using incentives and disincentives. The best results will be obtained through E.E.C. agreements aimed at raising standards in order to make research into new technologies and their adoption or the conservation of resources more profitable.

As far as products are concerned, it is best to use taxes on manufactured goods to discourage use of products which damage the environment either during or after use. Another possibility involves rigid analysis of new and potentially harmful substances during their introductory phase. Gradual limitations or eliminations are inevitable.

Activities aimed at spreading information about non-polluting technologies, as well as other information for consumers, could also be very useful.

Environmental Policy as a Means of Analysis and Check Lists of Specific Policies

Talking about the appropriate level of action does not mean that the real heart of the problem has been tackled. In fact, the real problem lies in the verticality of decision-making in public administrations, both on the local and national level. A good part of the policies mentioned above do not fall within the province of the Minister of the Environment or his regionally-based counterparts, but rather belong to various state or regional ministries and sectors. Environmental policy must first learn to be a means to verify the other policies (e.g., industrial, transport). Therefore, it is necessary to compile check-lists relating to the various fields of activity which would make it possible to interpret aspects of proposals and projects from the point of view of specific environmental objectives, as well as what is missing and should be inserted.

There is no short-cut. A policy that provides for incentives in terms of energy conservation on the one hand and massive projects for the development of electric productive capacity on the other does not make sense. Nor does a law in favour of "biological" agriculture proposed at the same time as the usual incentives for traditional agricultural production make sense. What we must ask ourselves is whether the energy plan includes the reduction of environmental pollution resulting from the production and consumption of energy, or whether the agricultural policy envisions the reduction of consumption of toxic substances.

A plan for the environment should be a clear statement of the principal problems to be confronted and a courageous attempt to explain where to start, since where to stop is probably not yet clear to anyone. In any case, the important thing is to try to follow the right path.

References

Bresso et al., (1988). La tutela dell'ambiente in rapporto alla distribuzione della popolazione, delle infrastrutture e delle attività produttive. Situazione, tendenze e problemi (Environmental Protection in Relation to the Distribution of the Population, Infrastructures and Productive Activities. The Situation, Trends and Problems), Quaderni CNR, University of Ancona (see also the indepth reports on sample regions and the basic reports from the study, which are forthcoming).

I.R.E.R., (1983). Usi e consumi di suolo urbanizzato in Lombardia (Uses and Consumption of Urbanized Land in Lombardy), Milan.

Ministry of the Environment-Econpubblica, (1985). La spesa per l'ambiente in Italia (Italian Expenditure on the Environment), Rome.

Ministry of the Environment, (1987). Nota preliminare alla relazione sullo stato dell'ambiente (Introductory Note to the Report on the State of the Environment), Rome.

Chapter 15

PHYSICAL PLANNING AND ENVIRONMENTAL PROTECTION IN THE LONG TERM

H. Leeflang

The Relationship between Physical Planning and Environmental Protection; a General Overview

In The Netherlands only in the recent years the relationship between physical planning and environmental management has become the subject of systematic regulation and planning. The mutual relationhsip between physical planning and environmental protection, however, is centuries old. A good example of this is the bye-law which was issued by the City of London after the Great Fire of 1666. In this law it was determined that in future large fires could only be lit outside the city. This was in fact an early example of 'risk policy'. This example indicates the essence of the relationship between physical planning and environmental protection; physical planning allocates space to functions in society. Some functions are sensitive to environmental pollution. It is the task of physical planning to draw up plans in such a way that the functions are so arranged that they do not influence each other in a negative way, but also support each other in a positive way (interweaving). The task of environmental policy is to establish norms of environmental hygiene and to ensure that pollution is tackled primarily at source.

The first legal regulations which enabled the Dutch government to pursue an environmental hygiene policy date from 1815 (the Nuisance Act) and 1901 (the Housing Act). The Housing Act also formed the basis for legal regulations in Dutch physical planning. Around the turn of the century the living conditions of the poorer inhabitants of the industrial cities in the Netherlands were very bad. It is therefore not surprising that in the preparations which led to the Housing Act, environmental hygiene had such an important role; housing is, after all, an environmentally sensitive function. The Nuisance Act was oriented upon environmentally harmful functions: it was concerned with 'the prevention of danger, damage and hindrance' from industrial activities to man. The Nuisance Act and the Housing Act can be seen as the basis of physical planning and environmental policy. As in many other countries, physical planning and environmental policy followed different lines of development. In The Netherlands physical planning became a central political issue in the 1960s. In the 1980s environmental awareness has grown and environmental policy has become more important. Physical planning and environmental protection (partly because of this development) have acquired rather different characters. Physical planning: complicated procedures, long reports, dull, very conservative, integrated. Environmental protection: trouble shooting, preparation for action, fully developed, little tradition, aspect-oriented. In the descriptions of physical planning and environmental protection as we see it in The Netherlands, there is an important common element present: striving for a satisfactory quality of living environment for man, animal and plant. If we further analyse the concept of quality as it is interpreted here it becomes clear how closely related both policy fields are. We can state that the quality of the living environment is determined by the values which we attribute to it:

The functional value of the environment and space:

* refers to the optimal (or sub-optimal) location of functions in space for different land uses; this is on the one hand determined by spatial planning and policy and on the other hand by the environmental hygiene situation of a certain area. In this way the prevention of soil pollution or the proximity of sensitive functions contribute to the functional value of land.

The perception value of the environment and space:

* refers to how man feels in his (built) environment; the size and scale of the built environment must be related to the well-being of people; for relaxation and recreation, the atmosphere in the rural areas must be acceptable: a task for physical planning. This value is, however, also heavily influenced by the environmental hygiene situation of the surroundings: a city full of polluting cars and industries is not pleasant, neither is a rural area with rubbish dumps and polluted water.

The future value of the environment and space:

* refers to return on investments in the (built) environment; here the environmental aspect comes very much to the fore since a great deal of environmental pollution is irreversible. For physical planning, which is in particular oriented upon future developments and use of space, it is very important that for long-term planning an adequate environmental hygiene quality can be depended upon.

These three elements of the quality of the living environment illustrate that an adequate physical planning policy is not possible without an adequate environmental policy. We call that a supportive environmental policy. With regard to the interdependence of functions in society, their locations and the associated environmental quality, the need from the viewpoint of environmental policy for a satisfactory physical planning policy can also be expressed: we call that a supportive physical planning policy.

What is the contribution of modern physical planning to a satisfactory environmental policy in concrete terms? We deal here with three different aspects:

a. **regulation of activities;**
b. **physical buffers (the zoning instrument);**
c. **assimilation in plans.**

a. In The Netherlands physical planning has a legal basis, the Physical Planning Act. In this act certain powers are allocated to public authorities at three levels (national, provincial and local) for land use control. At these three levels it is possible to take account of environmental aspects via spatial plans and regulations. Through a system of permits, environmentally harmful activities can be prevented from locating too close to environmentally sensitive functions. But it is at the local level that the legally binding decisions are laid down.

b. In The Netherlands environmental policy has two approaches: the source-oriented approach and the effect-oriented approach. In principle, pollution is tackled at source via the source-oriented approach. Where this does not have sufficient direct effect, the effect-oriented approach is utilized. By setting limits to the environmental nuisance of certain sensitive functions, it can be calculated what physical buffer must be maintained between sensitive and harmful functions. This we call environmental zoning. Within such a zone restrictions are set on the type of land use, installations and buildings. In physical planning we have an ambivalent approach to the zoning instrument. In the positive sense, this measure produces short-term results in the sense that set norms can be adhered to and firms are informed about the allowed level of environmental nuisance.
In the negative sense, it is in the first place a relatively easy solution to shift the blame of environmental problems onto physical

planning, and thereby to weaken the source-oriented approach; secondly, the total indirect land use taken up by environmental zoning can be extremely large, whereby the necessary flexibility in physical planning is restricted. In The Netherlands in due course there will be 2 to 3 million hectares of land use indirectly accounted for in this way. In our opinion, it is time, in the interests of a satisfactory physical planning, to evaluate the costs and benefits of source-oriented and effect-oriented measures more clearly. It is important to note that environmental norms are centrally set but that spatial quality is evaluated at the local level and is often neglected.

c. In The Netherlands over the past five years a new planning system for environmental policy has been produced. This system is based on the physical planning system. That does not mean that it is a perfect mechanism but in comparison to other countries, we are proud of our 'planning machine'. The strength of the whole system is that a large proportion of policy implementation rests with the regional level (provinces). A large measure of the relationship between spatial problems and environmental problems can be observed at that level, for example waste disposal, traffic measures, industrial estate planning, water management, housing programmes, etc. There are many possibilities for relating plans and implementation programmes and in future years they will be extended. The contribution of physical planning to a satisfactory environmental policy at the regional level is the setting of appropriate priorities for prevention and reconstruction within partial areas, the designation of high value nature areas and so-called protection areas for special functions such as drinking water acquisition and ground water protection as well as relating to it the location of other functions such as agriculture and livestock farming. As an illustration of the need and possibility of a supportive physical planning policy, in this paper three current examples are presented:
- the example of intensive livestock farming;
- the example of mobility;
- the example of the urban living environment.

Three Areas of Mutual Support

(1) The example of intensive livestock farming

Since the foundation of the European Community intensive livestock farming (pigs, poultry, cattle) has been able to develop explosively in The Netherlands. Although The Netherlands is small (20,000 km^2 agricultural land) and densely populated, it is one of the greatest agricultural exporters in the world. 50% of the production of intenstive livestock farming is exported. Imported into The Netherlands is cattle feed which is produced on an area two or three times as large as The Netherlands. A large proportion of this feed is used in intensive livestock farming which is concentrated on the sandy areas in the South and East of The Netherlands. In these areas various problems exist.

First, there is the problem of fertilizer surpluses; a surplus of 27 million tons above what is needed for the land. There are many municipalities with a combined average surplus of 4500 kg of phosphates over all cultivated land. Fertilizer surpluses have amongst other things a strong negative influence on drinking water and nature, especially brooks and streams.

Secondly, there is the problem of ammonia which emanates from the fertilizer. This ammonia is responsible for 30% of the acid rain in the Netherlands. In the sandy areas ammonia from animal fertilizer accounts for 30 to 80% of the acid rain. Large wooded areas, which are situated precisely in these sandy areas, will be lost.

Thirdly, there is the nuisance of smell. It is becoming more and more clear that the nuisance from smell is not just a temporary or local inconvenience but in whole regions leads to an increasing incidence of sicknesses such as chronic bronchitis and headaches.

Finally, there is the direct and indirect damage to the landscape. The landscape is damaged directly by intensive building activities in the rural areas. In the concentration areas on every hectare more than 100 m^2 is built up. The landscape is damaged indirectly by the disappearance of small landscape elements and the lack of variety signified by the ubiquitous nature of maize, the only crop which can cope with excessive fertilizer.

In physical planning the problem of intensive livestock farming was already recognized at the beginning of the 1970s. At the national level only a very global policy was formulated which implied that non-land based agriculture should be discouraged from rural areas. At the provincial level this policy was less rigid. Farmers' interests are often very well represented at this level. Moreover, it was apparently not easy to legally distinguish intensive livestock farming from other types of animal farming. The actual decisions are taken at the municipal level; here farmers' interests are apparently so dominant that intensive livestock farming can grow virtually unrestricted.

It is only in the past few years that a proper distinction has been made between intensive livestock farming and other agricultural activities. It was possible, however, to keep zones around villages and nature areas or very vulnerable agricultural areas free from intensive building activities. Topdressing of these areas can, however, hardly be prevented by spatial plans. At the provincial level more attention is now being given to intensive livestock farming. Real restrictions to growth are, however, not in operation. In the environmental field, the Soil Protection Act and the Nuisance Act are important (in addition to the administrative Fertilizer Act which falls under the auspices of the Ministry of Agriculture). The Nuisance Act is the oldest of these laws. It was oriented upon the prevention of nuisance to man. Only much later came the most important laws for water and air. These attribute the problems partly to the land, the soil. The Soil Protection Act has been passed after years of deliberation. This act regulates the following with regard to the fertilizer problem. The peaks in the environmental damage will be gradually removed. Around the year 2000 the whole of The Netherlands must have reached a level where fertilizer pollution has been reduced to an 'acceptable' level. That is still a level which is associated with disastrous effects on nature and drinking water.

Meanwhile the results are very limited. The growth of intensive livestock farming has now virtually halted. Only in empty (=clean) areas growth is possible. In the Land Reconstruction Act no account is taken of differences in function or the vulnerability of the soil. Soil and drinking water protection areas can, however, be designated on a very limited scale. This development is from the viewpoint of physical planning, not favourable in all respects. With a view to the maintenance of variety and the demands of the various functions and areas a differentiated set of norms would be desirable. The possibility of designating special protection areas is served here in only a very limited way.

In the long term it is hoped that the differentiation is realised. In the Nuisance Act there is a widening of objectives. The renewed Nuisance Act has now been in operation for a number of years. With this renewed act it has become possible not only to take account of nuisance to man but also the effects on nature and landscape. The ecological guidelines can in this way function as the basis for zoning around vulnerable areas.

Some conclusions from the above:
It is clear that environmental and physical planning are growing closer together and need each other to achieve optimum results. Norms are now guiding functions with the danger that a dispersal of functions occurs which can lead to a lack of variety. Functions should, therefore, also guide norms.

This is especially the case with more diffuse environmental problems at the regional scale. For larger emissions source-oriented and spatial policy can together achieve satisfactory results. For more diffuse pollution spatially differentiated environmental norms must be developed. The question remains, however, with such norms whether priorities are given to the protection of the most vulnerable or clean areas or the reconstruction of the most polluted.

The perspective for the mutual strenghtening or even integration of environmental and physical planning does not only comprise the development of norms differentiated by area and function. An active role is also expected from physical planning. In the context of national physical planning policy proposals have been made to realize differentiation within the sandy areas. On the one hand, additional instruments for the protection of vulnerable functions are introduced in landscape and ecological areas of special value. On the other hand, proposals have been made for the renewal of landscape, agricultural, ecological and environmental technical infrastructure in the central areas of intensive livestock farming.

(2) The example of mobility

Everyone who is involved with physical planning and environmental protection comes up against spatial, economic and environmental hygiene problems caused by the car. The intensity of physical mobility in a highly organized society can be regarded as one of the barometers of the economy. Revival and decline have a direct effect on the number of movements for all kinds of motives (commuting, business). It is, however, hardly possible to make a satisfactory cost-benefit analysis of what car mobility means in financial terms for society because benefits cannot be estimated. On the cost side, the finance which the authorities devote to counteracting environmental pollution by the car is very considerable. In The Netherlands, these costs will rise to about 300 million dollars per year around the year

2000. The costs to society of environmental pollution by car which are concerned with damage to health, flora and fauna, the built environment, agriculture and landscapes are many times in excess of this.

The Netherlands has the highest concentration of automobiles in the world: 128 per square km. This is a result of the population concentration and concentration of firms. In this respect The Netherlands can be compared to the Ruhr area or California. Since 1970 the road network has increased by some 15 %, whilst the traffic intensity has doubled in the same period. The road network within the built-up area increased by around 40%. All kinds of source-oriented measures have ensured that environmental pollution per car has decreased considerably with respect to certain substances. We expect a high increase in car ownership and car use, especially under the influence of economic growth, the process of individualisation and an increase in car ownership by women. The strong increase will ensure that the envisaged total decrease of emissions will be cancelled out.

In the counteraction of an unbridled growth of car mobility and thereby environmental pollution, physical planning plays an important role in the long term. In the long term a careful adjustment of especially housing and employment locations will no doubt have a favourable effect on the level of car use. The example of the American suburbs in the 1970s and the early 1980s where firms suburbanised, unplanned and on a large scale is so alarming in its consequences for traffic management, that there are enough arguments in The Netherlands to control the spatial organization of housing, employment and services.

In the Fourth Memorandum of Physical Planning, therefore, attention is devoted to the development of a spatial traffic and transport policy which has the objective of differentiating firms according to a mobility profile: a categorisation into the type of mobility which a certain firm generates at a certain location, divided into commuting and business traffic. Via a formula it can be determined from the viewpoint of traffic management and public transport use what the best location is for a firm. On the supply side, the location of firms should be planned in such a way that the traffic management is also guaranteed. That is what the authorities must do to guide the location of firms.

The role of public transport is important. Many American cities now face the consequences of a systematic neglect of the public transport system, in contrast to, for example, cities like London and Bonn where public transport takes care of a large proportion of the commuting traffic. As far as this is concerned, The Netherlands is lagging behind.

That does not mean to say that public transport is the only solution for the problem of the level of car use. Under the influence of all kinds of socio-economic developments, for example, the number of people with a fixed employment address is decreasing. In addition, people are less and less inclined to move to the place of employment, for example because they have a working partner.

All things considered it will be particularly difficult for the authorities to control de development of mobility. The contribution of physical planning to counteracting non-commercial traffic must, in the long term, lead to a reduced necessity for car use and thereby environmental pollution.

At the micro-level too satisfactory physical planning can lead to a reduction of environmental hygiene problems. By improving the traffic circulation and planning traffic-free residential areas both the perception value and the functional value of the environment can be increased.

(3) The example of the urban living environment

The relationship between physical planning and environmental policy is seen most clearly in environmental zoning. In the form of a physical zone a spatial distinction is made between environmentally harmful activities and environmentally sensitive functions. The best known are the so-called noise nuisance zones around airports, industrial areas and roads. By environmental zoning the effects of environmental pollution are in the first instance counteracted. Pollution sources are not tackled directly by zoning because it indicates - for example in local and regional plans - which functions (housing, employment, recreation) are best suited to which locations. On the basis of a certain pollution level, quality demands are made on a sensitive function such as 'housing' in the form and size of the zones. By using an overlay technique with the computer it can be determined in a simple way which functions are covered by the zones. If we have adequate information about the underlying spatial elements such as housing, schools, old people's homes, etc. then we can analyse the consequences of a certain zone or awareness of quality for physical planning on the one hand and the reconstruction programme of environmental policy on the other. This is important at the moment that strategic decisions are taken about spatial development in a municipality. From the viewpoint of efficient land use and spatial quality it is important to keep the environmental zones as small as possible.

In the municipality of Maastricht an experimental integration of spatial planning and environmental planning has been carried out as far as that was related to environmental influences which could be zoned. The counteracting of environmental pollution via a physical buffer is only concerned with certain forms of environmental pollution, namely noise nuisance, external safety, smell and other local forms of air pollution. The experiment was not related to the so-called protection zones around drinking water wells etc. It is clear from the project that in the approach to integration between physical planning and environmental planning, scale plays an important role. It is important to take account of the scale at which environmental problems and their possible solutions take place. Often the scale of the local plan is too small to make a satisfactory spatial decision about the location of harmful and sensitive functions. As far as environmental nuisance from industrial activities and from road traffic is concerned the whole area of the municipality must be taken into consideration. For nature functions and, for example, the quality of water it is even necessary to act at the regional level.

Working method in the municipality of Maastricht.

In order to achieve integration the emissions from stationary and mobile sources within the municipal area must be noted and, on the basis of existing environmental norms, incorporated into zones on maps. Parallel to this, ideal spatial pictures are drawn up by the municipal planners. In addition, the municipal area is divided up and described in outline according to existing and desired functions. Next, the results are confronted with each other and it can be determined in which areas the quality demands were not met on the basis of the environmental norms.

Ideally this confrontation should have an iterative character whereby each time a further evaluation of the desired spatial structure and the level of environmental nuisance in different intensities of reconstruction should take place. The result of such an exercise is a list of moments when a responsible evaluation should be made of the desired spatial structure and the possible reconstruction of the environmental pollution. It could, for example, be decided to:

- designate certain industrial areas as 'clean';
- indicate new locations for heavy industry;
- remove certain pollution sources;
- change traffic regulations;
- change the location of sensitive functions.

By working in this way it can be determined for each area which sources are responsible for the actual environmental quality and a reconstruction programme can be drawn up.

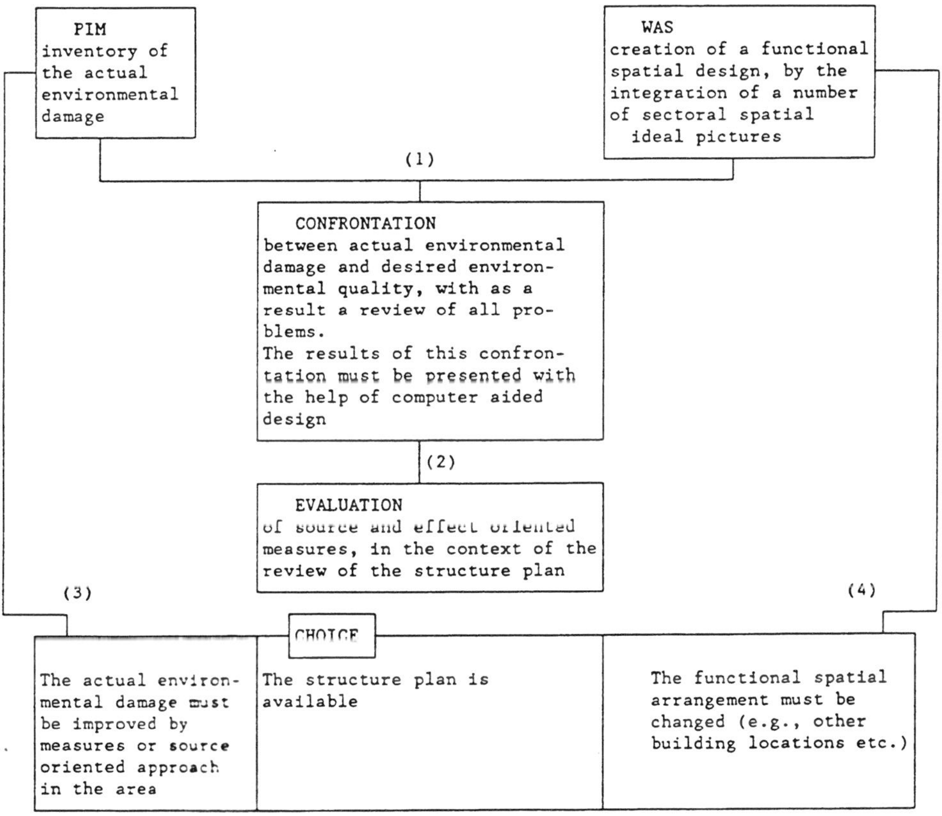

Figure 1. Schematic representation of the working process.

A More Fundamental Approach

The three examples show that action must be taken on a number of fronts at the same time in order to be able to guarantee the functional and perception values of space, also in the future. A close harmony of physical planning and environmental policy is necessary. The question is, however, whether the indicated policy will be sufficient. In the example of mobility it was apparent that the possible reduction of harmful substances per car will be cancelled out by the growth in the number of cars. Is a more fundamental approach possible?

On the way to the 21st century we, in the developed countries, will be confronted with an increasing use of raw materials and as a consequence of that increasing waste. This does not only result in an increasing dependence on other countries and associated with that an increasing vulnerability of the industrial society, but also in a more and more far-reaching and irreversible environmental pollution. As is apparent in the report of the World Commission on the Environment and Development there is reason for alarm and at the same time there are starting points for a fundamental approach to the problem.

As far as the use of raw materials and the production of waste are concerned, three important developments are awaited: **increasing production and consumption, technological innovations and a growing awareness of quality.**

More production and consumption could lead to more waste of raw materials and energy and more pollution. In the future the tap will not be opened so thoughtlessly. It no longer goes without saying that there is enough drinking water for consumption or that polluted water can be released. The Netherlands is a country with a great deal of water but is threatening to become a country where clean water is scarce. Of the total area of ground water, around the year 2000 8.5% is threatened by nitrate pollution. Surface water is already equally threatened by various forms of pollution. In addition the quantities of waste will rise considerably if no measures are taken. As a consequence of the - in themselves necessary- measures to tackle pollution at source, more waste is produced for which space must be found. Waste substances which are produced as a result of the prevention of emission into the air, water or soil have a large volume and can hardly be accommodated. The available dumping space in The Netherlands is very limited.

Technological innovations will create the possibilities of a more efficient use of raw materials, to utilise alternative raw materials and, more than is now the case, to convert part of the waste into new raw materials. The economic value of certain waste products is considerable: re-use of building material waste in The Netherlands already produces an annual saving of around 30 million guilders, and a similar figure applies to waste paper. With regard to industrial waste and non-ferrous metals the annual saving is even greater: in both cases around 400 million guilders. Re-use of 'bulk waste products' such as building material waste reduces the need to use more land for acquiring surface minerals such as sand and gravel. In the waste disposal system technical possibilities are available for a more environment-friendly, efficient and cheap system, pollution to be tackled at source, harmful substances to be removed from waste and the volume of waste to be reduced. These possibilities are also available for water purification. Technological innovations enable greater development in the use of wind energy.

The growing awareness of quality amongst the Dutch means that greater demands are made on the functioning of cities, and, increasingly, rural areas and the large water areas. In the past, the rural areas were the source of many raw materials and the dumping ground for much waste. One is more and more conscious of the need to maintain spatial quality in those areas. If waste and pollution become dominant a country like The Netherlands becomes more dependent on other countries for the import of raw materials and the export of waste. If that suddenly would not be possible, for example because of a political situation or a natural disaster, the whole system would collapse.

A less vulnerable, less wasteful and less polluted society is possible if human activities are better related to the carrying capacity of the environment. Via an ecological approach to urban and rural areas human intervention can be restricted. In spatial and environmental policy this requires an area-oriented approach in which urban and rural areas must be seen as 'households' and not as separate systems. That requires at the same time much more attention for (material, energy and information) flows: what comes in to an area, what passes through and what comes out?

In the physical planning policy of the Dutch government, a perspective of change is oriented upon careful management of raw materials, energy and waste in relation to the available spatial qualities. Also in the strategy of the World Commission on Environment and Development oriented upon 'sustainable development' the maintenance and strengthening of natural resources and a re-orientation on technology and minimisation of risk is central.

The viewpoint of physical planning requires an approach which focusses on the region and its spatial possibilities. The level of urbanisation and thereby also the nature of problems, such as use of raw materials and drinking water is per region so different that only such an approach can do justice to it. Moreover, spatial plans offer a good platform for the mutual adjustment of different sector plans. A regional approach offers, moreover, the best possibilities to use the intrinsic resources of a region as frugally and efficiently as possible, to limit waste flows and to promote re-use of waste. In this way the region will spatially be less dependent on developments elsewhere.

In the regional approach to the purification of polluted water and to the use of certain intrinsic raw materials and energy sources, a more concentrated or a more dispersed approach can be chosen. To make the choice between concentration and dispersal, in addition to the characteristics of raw materials or the type of energy source, the spatial characteristics and qualities of a region are very significant. In very densely populated areas there are few other possibilities then concentration of acquisition and assimilation. In many other regions sufficient space is available for a dispersed approach. Water purification, for example, can take place in a concentrated, large scale form. However, such an approach demands mostly a great deal of energy and high costs whilst few possibilities exist for a multi-functional use of space. If more space is available there are possibilities to opt for a more dispersed and small-scale approach. Such systems demand more space and time, it is true, but cost less energy and less money than a concentrated approach.

These systems can play a role in the purification of water which, for example, is used for drinking purposes or is introduced into nature areas and polders, as well as in the purification of waste water. Moreover, these marshy areas can acquire a function such as nature or recreation area. Recent experiences teach us that a combination of water purification, nature development recreation and landscape form is possible. In The Netherlands such water exploitation areas can possibly be combined with measures to counteract soil subsidence and salinity.

As far as the utilization of regional energy sources is concerned, attention in The Netherlands is focussed on wind energy. In the use of wind energy, experience has been gained so far only with dispersed windmills. Experiments are meanwhile being carried out with large, concentrated forms.

A choice between dispersal or concentration is not necessary with regard to the assimilation and storage of waste. Dispersed assimilation has the risk that re-use possibilities will not be insufficiently utilised and disposable or re-usable waste products will be dumped. In physical planning policy, therefore, a concentration of the assimilation of waste must be aimed for. Only then the space use can be kept as small as possible, harmful substances can be removed and the emission of waste substances can be prevented. For non-assimilable waste and 'bulk waste products' which can only be assimilated on a large scale, an approach above the regional scale is desirable. This concerns for example the building of large burning and separation installations and the construction of very large dumping grounds.

The outlined perspective is in general based upon:

a. a regional approach where possible;

b. more attention for the use of intrinsic (natural and human) resources;

c. limitation and re-use of waste.

The population will thus be confronted with the resources which they use and with the effects of their production and consumption. This will often require a change in the existing spatial organisation. For physical planning this means a new dimension to the concept of 'efficient land use': how can activities be arranged in space that waste from the one activity becomes the raw material of the other?

This is not a plea for regional autonomy but for a more harmonious adjustment of human activities in nature so that dependence, waste and pollution can be limited. This all fits into the motto: **think globally, act locally.**

The author wishes to acknowledge the comments on an earlier version of this paper by Hans ter Velde, Bram van der Klundert and Dave Hardy.

Chapter 16

FOUR IMPORTANT CHANGES IN THE AMERICAN APPROACH TO ENVIRONMENTAL REGULATION

L.E. Susskind

Prologue

The United States has had a comprehensive system of environmental regulation for almost twenty years. Beginning with the passage of the National Environmental Policy Act in 1969 (and the creation of the Environmental Protection Agency in 1970), federal, state and local governments in America have been involved in activities aimed at eliminating air, water, noise, and other kinds of pollution and managing the country's natural resources more wisely.

During this twenty year period, there have been several important changes in the American approach to environmental regulation. We have shifted from (1) strict to flexible deadlines and standards; (2) independent forecasting of impacts and risks to integrated management of impact and risk reduction efforts; (3) purely public to expanded private responsibility for enforcement; and (4) an over-emphasis on costly and time-consuming litigation to a more balanced reliance on consensual approaches to resolving regulatory disputes.

While the American approach to environmental regulation will surely change again in response to political, economic, technological, and scientific pressures, it may be helpful for other countries to look closely at the way the American system of environmental regulation has evolved thus far.

Deadlines and Standards[1]

In the early 1970s the US Congress sought to reduce the risk that EPA would abuse its discretionary authority or succumb to pressure from regulated industries. It did this by spelling out goals for the agency to meet by specific deadlines. The 1970 Clean Air Act required EPA to achieve clean air by 1975, and the 1972 Federal Water Pollution Control Act required EPA to eliminate discharges into the nation's waterways by 1985.

Under the 1970 Clean Air Act, EPA was given three key deadlines:

(1) Thirty days after passage of the act, EPA was required to propose national ambient air quality standards to protect public health

(primary standards) and welfare (secondary standards).

(2) Within a year after proposing these air quality standards, EPA was required to approve state implementation plans prescribing specific emission limitations for various types of polluters. (The states had the authority to use transportation control measures, if necessary.)

(3) EPA was required to implement emission levels for motor vehicles that would reduce auto emissions by 90% within five years. (They were also required to implement emission levels for new and modified stationary sources of pollution and for pollutants the agency considered toxic).

Under the 1972 Water Pollution Control Act, EPA was given two more deadlines:

(1) Within one year after the Act was passed, EPA was required to promulgate effluent guidelines that designated allowable discharges for various industrial categories (such as cement manufacturers, petroleum refiners, etc.). The effluent guidelines specified best practicable technology (BPT) objectives that were to be met by 1977 and best available technology (BAT) goals to be met by 1985.

(2) Within two years after the act was passed, EPA was required to issue permits to individual manufacturers that would achieve PBT and BAT in the time allowed.

These were not the only laws requiring EPA to meet specific deadlines and standards. In 1970 and 1972 Congress expanded EPA's responsibilities for regulating the use of pesticides and ensuring more effective disposal of solid wastes. Congress also passed the Noise Control Act of 1972, the Safe Drinking Water Act of 1974, and the Toxic Substances Control Act of 1975 giving EPA additional objectives to meet.

The agency was not able to meet any of these specific timetables. As a brand new organization, EPA had to work out its internal administrative arrangements. Then, it had to initiate the basic research needed to provide a scientifically-sound basis for setting standards. The agency's authority was challenged in court and it took more than five years to legally affirm EPA's mandate.

As Alfred Marcus, an expert commentator on the history of EPA points out:

> Economic, technological, and administrative obstacles prevented the agency from achieving statutory goals according to the original timetable(s) established by Congress. These obstacles, however, were

not the only ones EPA faced in its efforts to make rapid progress in a short time. The media, professional norms, the courts, and interest groups influenced the agency's ability to achieve goals by statutory deadlines.

By 1977, over 90% of America's water polluters had achieved best practicable technology. And, by 1975, 75% of the major stationary sources of air pollution in the country complied with state implementation plans. In the year Congress intended clean air for the nation, about 30% of the nation's 247 air quality control regions met clear air goals. While the deadlines slipped, real progress was made.

One of the reasons Congress enacted strict deadlines and standards was to ensure that America's national environmental protection goals would be achieved quickly. Most pre-1970 legislation had vague regulatory objectives. Administrative theorists argued for strict standards and deadlines as a remedy to lax regulatory enforcement. But when strict deadlines and standards also failed to achieve what Congress intended, there was a move toward still another set of regulatory reforms - 'market incentives'- highlighting greater flexibility.

Under the 1977 amendments to the Clean Air Act, for instance, EPA was allowed to create a limited 'market' for emission rights in non-attainmont areas. Basically, this allowed a factory in an area that had not met its clean air objectives to expand by buying the 'right to pollute'. It could do this by purchasing and closing down an old polluting plant in the same area while expanding its existing plant. Thus, instead of holding every facility in the same industrial category to the same strict technological and emission requirements, EPA could be more flexible as long as it met area-wide performance standards.

Advocates of 'market-incentive' approaches to regulation argue that EPA has not yet gone far enough in exploring still more flexible approaches to environmental regulation. These advocates are fond of pointing out that the 'heavy hand' of government regulation is inefficient and not cost-sensitive enough. For example, they cite various studies suggesting that the cost of achieving 95% effluent reduction in the United States will be around $119 billion while the cost of achieving the remaining 5% reduction (to get to 'zero discharge') will add $200 billion more. This, they conclude, proves that arbitrarily set administrative standards are usually less appropriate than market-driven approaches to regulation.

No one is quite sure whether a more flexible approach would have achieved as much as strict standards and timetables achieved in the 1970s. Also, since it is difficult to measure the full range of costs and benefits associated with each attempt at environmental regulation, it is hard to say whether flexible approaches are really better. What is clear, though, from the American experience to date, is that strict deadlines and timetables will always fall short.

There is much to learn along the way about how to implement each new environmental regulation. Mid-course corrections are necessary. Legal requirements alone do not compel the necessary technological innovation; these take time. Inflexible deadlines and standards ignore the need to take account of agency experience as it accumulates.

Impact and Risk Assessment

The National Environmental Policy Act of 1969 (NEPA) stipulates that 'for major Federal actions significantly affecting the quality of human environment' alternatives to each proposed action must be considered. The guidelines issued by the Council of Environmental Quality state that:

.....agencies shall:

(a) Rigorously explore and objectively evaluate all reasonable alternatives, and for alternatives which were eliminated from detailed study, briefly discuss the reasons for them having been eliminated.

(b) Devote substantial treatment to each alternative considered in detail including the proposed action so that reviewers may evaluate their comparative merits.

(c) Include reasonable alternatives not within the jurisdiction of the lead agency.

(d) Include the alternative of no action.

(e) Identify the agency's preferred alternative or alternatives, if one or more exists, in the draft statement and identify such alternatives in the final statement unless another law prohibits the expression of such a preference...

Federal and state agencies have responded to these requirements by developing procedures for evaluating alternative courses of action. Indeed, the art and science of environmental impact assessment has spawned new professional firms, journals, associations, and educational programs.[2] Many countries have adopted some form of the American approach to environmental impact assessment. But that approach has changed significantly over time.

As the practice of environmental impact assessment has evolved, it has become clear to close observers that non-objective judgements play a crucial part in what was thought to be an objective and purely technical forecasting activity. For instance, there are no technically correct geographic boundaries for circumscribing an 'impacted' area. Nor are there technically correct timeframes, discount rates, amalgamation techniques or indices of impact to use in preparing an impact statement. While these must all be specified before an impact assessment can be completed, expert judgement or techniques alone can not provide definitive answers.

EPA recognized this when the agency adopted new 'scoping' guidelines in 1977. The requirement to scope each impact assessment is basically a stipulation that the lead agencies (which must prepare formal impact statements) meet with all groups who might be upset about the non-objective judgements that will be made in the course of preparing the impact assessment. Presumably, the scoping process will identify acceptable non-objective (as opposed to technically correct) judgements.

Impact assessment requirements are nothing more than common sense 'look-before-you-leap' procedures. They presume that it is possible to forecast the adverse environmental impacts likely to be caused by a proposed action (or alternatives to that action). Initially, Congress presumed that independent analysis could be called upon to prepare technically correct forecasts that would then be scrutinized by all those who are part of the decision-making process. The appropriate decision-makers are expected to make the 'environmentally intelligent' decisions (i.e., choosing either the least environmentally detrimental alternative or accepting responsibility for mitigating the adverse environmental impacts associated with the course of action they have chosen). This, however, did not always happen. Many times, the EISs were ignored or prepared in a manner designed only to justify decisions that had already been made.

Most analyses of NEPA (and related state NEPAs) are not very positive about the extent to which impact assessment requirements have improved the quality of environmental decision-making in the United States. The best that can be said is that impact assessment requirements have probably caused 'really terrible' projects not to be proposed in the first pace. There are very few examples, however, of major projects or policies that were radically altered as a result of an impact assessment.

The most sophisticated environmental policy analysts in the United States now question whether it makes sense to separate impact assessment procedures from all other decision-making requirements. Indeed, few environmental planners would argue any longer that an independent analyst should prepare an impact assessment 'at arms length' from those involved in and affected by the proposed action under scrutiny. The participation of all those affected by a decision in every aspect of impact assessment design and implementation is now seen as the only way of ensuring that important non-objective judgements are made in an accountable fashion.

The same evolution is being repeated as risk assessment emerges in the United States. Risk assessment is yet another procedure that agencies responsible for public health and environmental quality are now expected to perform as part of decision-making. Initially, the thought was that risk assessments would be provided by independent analysts who would not be 'influenced' by political or other decision-making considerations. As it has turned out, though, risk assessment is also more of an art than a science and dependent on a series of non-objective judgements (such as the value that ought to be attached to human life or the appropriateness of extrapolating from laboratory rat assays to potential health impacts on human populations). In the past few years, there has been a substantial debate about the proper link between risk assessment and risk management. We seem to be moving away from the notion of independent analysis by experts to integrated decision-making by all those affected.

The notion that the link between analysis and action in environmental decision-making ought to be severed is probably wrong. What we have realized in the United States is that we need to make the link between the two explicit and learn how to manage the linkage more effectively.

Expanded Private Responsibility for Enforcement

As interest in environmental protection grew in the United States in the 1960s and early 1970s, public awareness of the lack of credible enforcement mechanisms increased. This awareness and the public dissatisfaction it engendered led directly to the creation of citizen suits.[3] The citizen suit requirements in federal environmental legislation are intended as a response to the government's failure to enforce compliance. Congress believed that citizen suits would provide a goad to government enforcement and, if that goad did not work, an alternative means of enforcement.

The citizen suit was included in almost all new federal environmental statutes or major amendments in the 1970s. Citizen suits 'authorize any person to commence legal action to enforce the requirements of an act against any person alleged to violate them or to require the government to perform a mandatory duty under the act.' They confer jurisdiction on the federal district courts to hear such suits without regard to who is bringing the action or the costs at stake. A suit may not be commenced if EPA is 'diligently prosecuting an enforcement action.'

While the citizen suit was conceived and designed to allow private enforcement of the law against pollutors, its most celebrated uses have been against EPA itself for its failure to implement its environmental statutes in a timely and complete manner. There have not been as many citizen suits as proponents expected. Probably fewer than two hundred suits have actually been heard to completion. Many of these were undertaken in the early 1980s when EPA seemed to be pulling back from its enforcement responsibilities. A growing number of notices of proposed citizen suits are filed with EPA each year, forcing the agency to act (and forestalling completion of the suits while EPA seeks compliance).

Citizen suits typicaly focus on failure by an implementing governmental agency to perform mandatory duties. Some also cover enforcement against violations of rules, regulations, permits, licenses, or leases.

Under the Superfund Law (passed in 1980 and amended in 1986) corporations and not just citizens are pursuing private enforcement actions. Companies that own property containing hazardous wastes are suing other companies who, while not currently connected with the wastes, historically were connected with them. They have begun suing for the cost of cleaning up these sites.

The availability of a private cause of action to enforce Superfund closely resembles the various citizen suit options contained in most pieces of environmental legislation.

The net effect of these citizen suits and corporate causes of action is to expand private enforcement. Unfortunately, this can result in substantially different degrees of cleanup than might have occurred at the same sites if the government were handling enforcement. Thus, private enforcement may result in under- or over-enforcement. Various efforts are underway to remedy this situation.

There has been a shift from purely public enforcement to expanded private enforcement of federal regulations. This is partly the result of EPA's unwillingness in the early and mid-1980s to vigorously enforce the law. Private enforcement received support from both citizen activists (who do not feel that government can be counted on to be vigorous enough in its enforcement efforts) and regulatory reformers (including business leaders) who view privatization of enforcement as a step back from an ever-growing federal enforcement bureaucracy.

Alternative Dispute Resolution[4]

There are several conferences every year in the United States devoted solely to reviewing the latest court cases extending and clarifying judicial interpretations of environmental statutes. The number of lawyers specializing in environmental litigation has grown substantially. For many citizen action groups and public interest organizations, litigation (through the citizen suits I have already described as well as through more traditional legal challenges to administrative actions or inaction) has been the primary means of pressing for enforcement of environmental protection statutes.

There has been a growing dissatisfaction, however, with the cost and time consuming nature of litigation. Moreover, the courts are increasingly unable to cope with the science-intensive nature of most environmental regulatory disputes. The courts, concerned about overcrowded dockets and complex cases that can tie up a judge and a jury for years, have endorsed the search for alternative dispute resolution (ADR) techniques.

The ADR movement in the United States is still in its infancy. Yet, numerous complex civil cases have been settled through informal negotiation and mediation while other 'hydrid' ADR techniques (including mini-trials, negotiated rulemaking, summary jury trials, and joint fact-finding) have been used to unknot enforcement disputes, licensing and permitting conflicts, and other disagreements that have not yet blossomed into full-scale litigation.

Several federal agencies, in particular EPA, have been experimenting with various forms of 'assisted negotiation' aimed at encouraging settlement without the cost and delay of litigation. While these experiments have for the most part been successful, they have not been without controversy. Agency personnel are often suspicious of and defensive about negotiated settlements. If they settle, they may appear to have 'sold out.' Furthermore, involving mediators or other 'neutral intervenors' appears to infringe upon agency perogatives. Agreeing to 'assisted negotiation' might even appear to be admitting failure.

Environmental advocacy groups have also been reluctant to back alternative dispute resolution too enthusiastically. In some instances, they need a clear cut court decision to establish 'whether' EPA has the right or responsibility to do something. When they want to set a precedent, negotiated settlement of a case will be of little or no help. Environmental advocates are also nervous about EPA negotiating 'deals' with industry groups. The environmentalists sometimes feel 'outgunned' in informal negotiations where technical expertise or financial resources are not distributed evenly. They have come to believe that only the courts provide a 'level playing field.'

ADR advocates are quite careful to spell out the circumstances under which negotiated settlement (whether assisted by a neutral mediator or not) makes sense and when it does not. They argue, for example, that constitutional issues and questions about fundamental rights should not be negotiated. But once these issues have been litigated, case-by-case application of general principles might better be handled through negotiations in which all interested parties are given whatever assistance they need to participate on an equal footing.

The Administrative Conference of the United States (ACUS) - a body made up of senior administrators from every federal agency - has recently endorsed the use of ADR techniques in environmental and other kinds of civil disputes. The endorsement by ACUS has given ADR new legitimacy.

The shift from litigation as the primary means of dealing with differences between EPA and industry or between EPA and environmental organizations to a search for alternative dispute resolution techniques is only just beginning in the United States. In a country as devoted to litigation and lawyers as America is, it is nothing short of amazing that consensual approaches to environmental dispute resolution are gaining any attention at all.

Summary

When the United States broadened and expanded its environmental regulatory activities twenty years ago, there was no clear logic to the design of the regulatory system that emerged. Through a process of trial and error, reforms have been proposed and adopted. What began with a focus on strict statutory deadlines and standards has evolved to a more flexible system, including partial reliance on market approaches to regulation. Independent forecasting of impacts and risks in advance of decision-making is more common than ever, but now instead of viewing forecasts as the responsibility of expert analysts whose technical judgements should not be protected from political and non-objective considerations, we think of impact and risk assessment as part of a larger managerial process of reducing adverse impacts through decision-making that acknowledges the importance of political and non-objective considerations.

What was once a regulatory scheme that presumed only public responsibility for enforcement has broadened to encompass citizen suits and other forms of private responsibility for enforcement. And at the same time as this has increased the importance of litigation as an enforcement mechanism, we have seen the rise of the ADR movement - a search for consensual approaches to negotiating settlement than will cost less, take less time, and truly reconcile the legitimate differences of regulators and regulatees.

While more needs to be done to perfect the American approach to environmental regulation, I believe these changes represent important steps in the evolution of the American system. They have not been taken as pruposefully as I may have implied. They have evolved a bit at a time. The transformation they represent, however, may well be instructive for other countries, particularly those that are now seeking to avoid some of the mistakes that the United States has made.

Notes

1. This first section draws heavily on Alfred Marcus' excellent history of EPA. 'Environmental Protection Agency', in James Q. Wilson (Editor), The Politics of Regulation, New York: Basic books, 1980.

2. The changing nature of the practice of impact assessment and risk assessment is monitored in the international quarterly, Environmental Impact Assessment Review, published at the Massachusetts Institute of Technology.

3. For an excellent legal summary of citizen suits see Jeffrey Miller's three-part series entitled 'Private Enforcement of Federal Pollution control Laws' in The Environmental Law Reporter, October-December 1983.

4. This section draws from Lawrence Susskind and Jeffrey Cruikshank, Breaking the Impasse: Consensual Approaches to Resolving Public Disputes, New York: Basic Books, 1987.

Chapter 17

ECONOMIC GROWTH AND ENVIRONMENTAL CHANGE IN JAPAN; NET NATIONAL WELFARE AND BEYOND

K. Uno

Introduction

The need for quantitative measurement of the state of welfare had gradually been accepted in Japan as controversy over growth had become intense. The question was raised whether economic growth contributed to the betterment of economic well-being or not, what contributions economic growth brought about to the quality of life, and what kind of policy helped toward the expansion of welfare. This led to the estimation of Net National Welfare (NNW) by the Japanese government. It was in May 1971 that an NNW Development Committee was established within the Economic Council. This Committee announced 'New Welfare Indicators: NNW' in March 1973. Prof. Shinohara of the Institute of Developing Economies (then, the Director of Economic Research Institute of the Economic Planning Agency) played a leading role in the Committee. The NNW actually is a Japanese label for Measures of Economic Welfare (MEW) proposed by Nordhaus and Tobin (1972).

Several years later in 1976, an updating of NNW estimation was attempted by Mr. Kanamori, Ms. Takase and the present author, who at that time were connected with the Japanese Economic Research Center. The results of their estimation were reported in detail in Kanamori et al. (1977).

Their analysis which covers the period up to 1975 is summarized in the next section of this chapter, together with the estimates for the 1980-1985 period by the present author. This section, therefore, brings the NNW analysis up to date by inclusion of the decade following the 1973 Oil Crisis. This period saw considerable effort at pollution prevention as well as rapid structural change of the economy. In our 1977 study it was concluded that environmental disruption was the largest hindrance in translating economic growth into economic well-being. It is interesting to see what has happened since then.

Next, an analysis is attempted to look into the factors which contributed to conservation of energy. Air pollution is closely related to the consumption of sulphur-containing resources, and conservation of energy naturally helped contain the emission of pollutants. As discussed in a final section, coupled with increased provision of pollution prevention facilities, pollution levels were improved considerably, resulting in less complaints against environmental disruption.

Economic Growth and Economic Welfare

The implicit assumption that economic prosperity would automatically improve the quality of life was long held by not only those who are in the economic profession but also people in general. However, a growing number of people began to realise that this was not the case. "Growth was an important economic policy objective to Japan until the first half of the 1960s. Almost no one had question about the contribution of economic growth to the upliftment of the nation's level of welfare. But, in the latter half of the 1906s, sharp criticism against growth arose. It was because environmental pollution, the destruction of nature, price inflation, and the devastation of public feelings has become conspicuous. Did growth contribute to the advancement of welfare, or did it lower it? It is hard to answer this question. There is no agreement of opinions about what is welfare, and to quantify the level of welfare involves numerous difficulties in terms of both principle and technique. But it is worthwhile to try to define welfare in a way so that welfare may be measured and to try to look into quantitative changes of welfare. To do so is to understand but an aspect of welfare in a wide sense of the word, but, without it, discussion of welfare would get nowhere and the evaluation of growth and environment policies would be impossible."[1)]

NNW is a modification of GNP, and includes only elements which can be somehow translated into monetary value. Basically, NNW consists of the following eight items:

(1) NNW government consumption;
(2) NNW personal consumption;
(3) services attributable to household-related social overhead capital;
(4) services attributable to personal durable consumption goods;
(5) leisure time;
(6) extra-market activities such as housewives' domestic services, etc.;
(7) (deduction) environmental pollution, and
(8) (deduction) loss due to urbanization.

NNW is derived as the sum of these items. Investment, which is an important component of GNP, is excluded from the calculation on the ground that investment is but 'intermediate goods' from the point of view of quality of life measurement and what directly contributes to welfare is the 'services' rendered by capital stock. NNW therefore explicitly accounts for the stock of consumer durables and household-related social overheads, and includes the imputed flow of services from them.

The estimated values of NNW in Japan are presented in Table 1, together with the values of GNP. It makes little sense to compare the absolute values of NNW and GNP and to point out that the welfare levels were lower than production levels because the former excludes investment. In fiscal year 1975 (which starts on April 1), NNW is calculated to be about 74,000 billion yen (in terms of 1970 prices). As for GNP, the 1975 value was about 93,000 billion yen. The largest component of NNW was personal consumption, followed by imputed value of leisure time and extra-market activities. Rapid growth of services arising from personal durable goods is also worth particular attention. At the same time, damages done by environmental pollution and loss due to urbanization had increased rapidly, pulling down the levels of welfare.

On the whole, the growth of NNW during the 1970-1975 period in real terms is calculated to be 9.3% on an annual average, compared to annual gains of 5.2% for GNP. In other words, the growth of welfare was maintained in the face of economic deceleration during this period due to restraining economic policy and stagnant private capital formation in the aftermath of the Oil Crisis. Higher levels of consumption, increased leisure time, larger stock of consumer durables and household-related social capital, and improvement in environmental pollution were the main contributors to the growth of NNW. Actually, average annual growth rate of NNW rose from 5.1% in the 1955-1960 period to 6.8%, 8.2%, and 9.3% in subsequent five-year periods.

The welfare loss attributable to environmental pollution hit the peak in 1970 in absolute number as well as in its ration to NNW, which amounted to 14,3%. By 1980, the loss caused by pollution was reduced to 4.3% of NNW; the 1985 figure is still lower at 3.0%.[2)]

We also note that the rate of growth of GNP is higher than that of NNW in the 1980s. This is a reflection of resumption of private capital formation in order to pave the way for information-oriented society incorporating computer-communication-control technology.

A summary description of the method of assessment is as follows:

(1) NNW government consumption: Government expenditures are classified by function and those for education and culture, social welfare, and health and sanitation are counted among components of NNW. Expenditures for judicial

Table 1. Net National Welfare (NNW)

(unit: billion yen, 1970 prices)

Fiscal year	1955	1960	1965	1970	1975	1980	1985
NNW gov't expenditure	1,199	1,374	2,254	2,988	3,865	4,283	4,887
NNW personal consumption	10,427	14,706	22,168	32,097	43,003	54,009	61,700
Gov't capital service	62	99	169	317	559	756	1,103
Personal durable goods service	91	195	755	2,342	4,187	5,270	6,813
Leisure time	4,871	6,098	7,325	10,509	16,759	18,961	20,816
Extra-market activities	1,876	2,388	4,068	7,213	12,707	12,571	13,079
Environmental pollution (deduction)	-38	-1,037	-3,735	-6,805	-5,729	-3,932	-3,103
Loss due to urbanization (deduction)	-452	-695	-889	-1,113	-1,119	-1,272	-1,514
NNW	18,036	23,128	32,116	47,548	74,231	90,646	103,781
GNP	17,268	26,183	41,591	72,144	93,260	118,105	143,387

Source: The figures for 1955-1975 are from Kanamori, Takase, and Uno, Economic Growth and Welfare---An Estimation of Japan's NNW---, The Japan Economic Research Center, 1977; those for 1980-1985 are estimates by the present writer.

and police services are excluded on the ground that they are 'defensive expenditures', having no net addition to welfare.

The items under this heading have exhibited consistent growth during the period under consideration.

(2) NNW personal consumption: Based on personal consumption as a component of GNE, this item is calculated after the deduction of the purchase of consumer durable goods (service of which is accounted for in item (3) below), commuting expenses, cost of obligatory payments including ceremonial services, etc.

Despite drops in the rate of economic growth, NNW kept growing due to personal consumption. Part of the explanantion lies in the fact that, unlike investment in plant and equipment, personal consumption continued to expand even in the recession years.

(3) Services attributable to government capital: Flow of services imputable to social overhead capital stock which are related to the well-being of the household are calculated. Examples are housing, city parks, schools, water supply, drainage, and waste disposal. The value of such services are calculated assuming depreciation rate of 6.5% over a twenty-five years life span.

Increased provision of such social capital had its effects on the imputed flow of services. In fact, annual rate of increase was considerably higher than that of NNW personal consumption; 13.4% vs. 7.7% in the 1965-1970 period and 12.0% against 6.0% in the 1970-1975 period. The absolute magnitude of such services is relatively small compared to other components of the NNW.

(4) Services attributable to durable consumer goods. The purchase of durable goods is recorded in the consumption in the current period in the conventional treatment. In NNW account, however, current purchases of durable goods are not counted in the current consumption, but the 'services' rendered by the stock of consumer durables. This reflects the benefit received by the household.

The number of washing machines, refrigerators, television sets, automobiles, and other durable consumer goods have literally exploded in the course of postwar economic growth in Japan, reflecting rising income levels and mass production of these goods. This was particularly so in the 1960s as can be seen from the growth rates of services rendered by such goods. For example, in the first half of the 1960s, annual growth rate of services attributable to durable goods is estimated to have reached 31.1%, and in the latter half, 25.4%. More recently, however, the growth of personal consumption centers around various personal and public services, and the annual growth of this item has dropped.

(5) Leisure time: Value of Leisure is imputed based on average wage per hour, adjusted by age group and by sex. The assumption here is that the wage rate indicates the value of leisure time foregone. The calculation covers the population who are fifteen years of age or older.

Leisure is defined here to include time spent for social activities, rest, reading, listening to the radio or watching television, and other personal activities. Average monthly working hours tended to be shortened reflecting people's preference for leisure over additional income. This was especially so since the mid-1960s when full employment was attained in the Japanese labour market and income began to rise consistently.[3)]

One problem here is that, in the recession years such as after the Oil Crisis, increase in leisure time includes an involuntary element of the reduction of working hours. Increased leisure in this case is in fact a negative factor in well-being. This results in overestimation of imputed value of leisure and of NNW itself.

There are three factors which affect the allocation of time. First is the time consumption pattern within a single day. In a weekday, for example, the Japanese spend 6 hours and 13 minutes in 1985 for leisure activities as defined in this chapter, which is about 20 minutes longer than in 1970, but is actually 4 minutes shorter than in 1975 and 11 minutes shorter than in 1980. This is largely attributable to the reduction of time spent for TV watching. Leisure activity in a narrow sense has steadily increased from 31 minutes in 1970 to 50 minutes in 1985. Second factor is the number of days off. Increasingly large numbers of people are enjoying two days off in a week instead of working half day on Saturdays, and the number of people who take paid vacation is increasing. In addition, national holidays have been added in recent years. The third is the age-sex composition of population, and the increased participation of women in the working force is changing the national time budget.

(6) Extra-market activities: This consists of the value of domestic services of housewives. The imputation is done based on the average wage of female workers and covers those who are fifteen years of age or older and who are exclusively engaged in domestic works. Housekeeping is not market transaction and is not considered to contribute to GNP, but it clearly contributes to well-being.

(7) Environmental pollution: As for environmental damages due to air and water pollution, and waste disposal, the estimated cost is based on the

required improvements to restore the pollution level consistent with the established environmental standards. Excess emission is first estimated and the removal cost is then calculated. Some items such as noise and bad odor are dropped from the calculation because of lack of appropriate data. In this sense, the estimated environmental damages understate the true value. Also excluded are the damages done to natural environment through various development projects and improper land use.

Estimated damages literally jumped from mid-1950s to 1970. According to the estimation, the annual rate of increase reached nearly 95% in the 1950-1960 period, and 29% in the 1960-1965 period. In the latter half of the 1960s, the rate of increase was estimated at 12.6%. It was only in the 1970s that the trend was reversed and for the period from 1970 to 1975, environmental damages declined at an annual rate of 3.4%.

(8) Loss due to urbanization: This item represents increased in commuting time and traffic accidents.

Commuting time tends to become longer in Japan and its negative contribution to welfare is estimated based on the average wage per hour. Among the persons with a job, national average time spent for commuting was increased from 43 minutes per day to 52 minutes in the five-year period from 1976 to 1981.[4)] In Tokyo metropolitan area, people spend 1 hour and 33 minutes per day for commuting, which is 37 minutes longer than in rural area.[5)]

In contrast, traffic accidents have peaked off around 1970, and the number of accidents actually declined. There are now fewer accidents and fewer deaths and injuries due to traffic accidents. As a result, loss of urbanization has not increased in recent years.

In the next section, we turn to a more detailed examination of the environmental aspect in het NNW framework.

An Estimation of Environmental Damages

The most detrimental to the quality of well-being is environmental pollution. It became a source of major social conflict in the 1960s and this is clearly reflected in the estimated trends of NNW.

According to Table 2 which shows the welfare loss due to environmental disruption in value terms, we observe rapid increase before 1970. The figures were derived by imputing the loss from pollution in terms of the cost of conforming to the environmental standards. In retrospect, we may say that, if investment for the expansion of production was partially diverted to pullution prevention at that time, gains in welfare must have been

Table 2. Loss due to Environmental Pollution

(unit: billion yen, 1970 prices)

Fiscal year	Economic Council estimates 1955	1960	1965	1970	Kanamori estimates 1970	1975	Uno estimates 1980	1985
SO2		289.3	750.7	1,797.6	1,727.9	1,167.9	668.7	459.8
Soot and dust	34.6	89.3	117.4	400.9	368.7	375.9	316.9	272.6
Automobile exhaust gas	0	0	269.8	1,151.2	1,032.9	1,906.2	1,176.1	973.3
BOD		557.7	1,142.8	1,505.6	1,733.3	1,481.6	1,213.4	997.8
Factories		299.3	630.7	832.0	1,068.7	639.5	---	---
Homes		258.3	512.1	673.6	664.6	842.1	---	---
Waste			1,035.4	1,245.6	1,942.0	797.5	556.9	399.0
Non-industrial			87.0	106.7	34.5	66.7	26.5	14.1
Industrial			948.4	1,138.9	1,907.5	730.8	530.4	384.9
Total	34.6	936.3	3,376.0	6,100.9	6,804.8	5,729.1	3,932.0	3,102.5

Note: NOX emission from fixed origins is not included here due to unavailability of data thereof. Figires for 1955-1975 are from Kanamori, Takase, and Uno, Economic Growth and Welfare---An Estimation of Japan's NNW---, The Japan Economic Research Center, 1977; those for 1980-1985 are estimates by the present writer.

greater even if at the cost of slower GNP growth. The trend was clearly reversed in the subsequent period. Loss from environmental pollution reached a peak in 1970 at 6800 billion yen (or about 54 billion dollars based on an exchange rate of 125 yen per one US$ which prevailed in 1988). This was reduced to 3100 billion yen in 1985 (1970 constant prices).

I have examined in a previous paper the formulation of environmental policy in Japan, together with the pollution prevention measures carried out in the private and the public sectors.[6] The discussion there was based on time series data, disaggregated to individual industries.

There are lots of information which are better dealt with within the framework of input-output analysis. In this paper, we shall examine these data in conjunction with the changing industrial structure. The cases in question are the pollutant emission accompanying various productive activities, such as air pollutants, water pollutants, and industrial wastes. As we have seen in the previous section, environmental disruption was extremely keen in Japan from mid 1960s to mid 1970s. The data used here were collectd around 1970 when the pollution problem was most threatening, not before the damages were actually done. In this sense, time-series comparability of pollutant output per unit production, for example, is rather limited.

Typical factors causing environmental pollution can be classified as follows

- Air pollution
  - Stationary sources
    - Sulphur oxides
    - Soots and dusts
  - Movable sources
    - Automobile exhaust (Nitrogen oxides, Carbon monoxides, Hydro-carbon)
- Water pollution
  - Industrial water
    - Biological oxygen demand
  - Domestic water consumption
- Wastes and garbages
  - Industrial wastes
  - Domestic wastes

Of these, we focus our attention on the ones which are related to industrial activity, i.e., air pollution from stationary sources, water pollution by industry, and industrial wastes. For each pollution category, we first examine unit output coefficients of pollutants. These coefficients give the amount of pullutant output in physical terms per unit of production activity; for example, several tons of industrial wastes per one billion yen worth of production. Second, we estimate the total quantity of pollutant outputs. This is achieved by multiplying the unit output coefficients by the amount of industrial production for each sector. Based on production statistics at various point in time, we can obtain the total quantitiy of pollutant emission. Third, based on the unit cost required to remove the pollutant, total removal cost is estimated. In the NNW framework, environmental damage is approximated by estimated expenses in order to properly treat the pollutants to meet environmental standards. Conceivably, there must be a way of directly assessing in money terms the damages to health, human life, and natural environment caused by discharge of these pollutants. Such 'direct method' was judged infeasible considering the availability of empirical data.

(1) Air pollution: As sources of air pollution, we can distinguish stationary sources such as factories and movable sources, namely automobiles. The air pollutants from the former include sulphur oxides (SOX), nitrogen oxides (NOX), and soots and dusts, among others. In recent years, dusts caused by spike tires which are used during the winter season have beome a problem. Also of a serious social concern is the nitrogen oxygen emission of automobiles. In this chapter, however, we focus on the stationary sources and confine our observation to sulphur oxides.

The quantity of emission of SOX by each industrial sector is estimated by unit energy consumption coefficient, which is obtained by dividing the energy consumption (Table 3) by the value of production. The social cost of air pollution due to SOX is calculated as the expenses necessary to remove the discharge in excess of the 1955 level.

Obviously, economic expansion brings about more enviornmental damages. The variables here are (a) the changing sectoral shares in the economy, (b) sulphur content in coal, heavy oil, and other raw materials, and (c) the removal rate of sulphur before emission.[7)] With due consideration to these factors, the damage attributable to SOX is estimated to have reached about 1700 billion yen in 1970 (Table 2).

Table 3. Energy Consumption in Base Year (1970)

Sector Number	Modified JSIC Code		Coal (unit: 10^3 ton) (S content 1.04%)	Heavy oil (unit: 10^3 kl) (S content 1.89%)
(1)	01&	Agriculture, forestry, and fisheries	--	4566
(2)	10&	Mining	1391	395
(3)	18	Food processing	264	3221
(4)	20&	Textiles	84	4637
(5)	22	Lumber and wood products	included in 38&	included in 38&
(6)	23	Furniture and fixtures	Included in 38&	Included in 38&
(7)	24	Pulp, paper, and paper products	584	5775
(8)	25	Printing and publishing	included in 38&	included in 38&
(9)	29	Leather and leather products	included in 38&	included in 38&
(10)	28	Rubber products	included in 38&	included in 38&
(11)	26	Chemicals	331	12517
(12)	27	Petroleum and coal products	1269	5095
(13)	30	Nonmetallic mineral products	234	10380
(14)	31A	Iron and steel materials	} 231	} 11107
(15)	31B	Basic iron and steel products		
(16)	32	Basic nonferrous metal products	—	2235
(17)	33	Fabricated metals	—	
(18)	34	General machinery	—	
(19)	35	Electrical machinery	—	
(20)	36M	Transport equipment, motor vehicles	—	} 4381
(21)	36S	Transport equipment, others	—	
(22)	37	Precision instrument	--	
(23)	38&	Other manufacturing	204	
(24)	15	Construction	--	808
(25)	70	Electric power, gas, and water	20182	34452
(26)	40	Wholesale and retail trade		
(27)	50&	Finance, insurance, and real estate dealing		
(28)	50R	Real estate rent		
(29)	60T	Transportation	604	4620
(30)	60C	Communication		
(31)	97	Public administration		
(32)	75P	Public services		
(33)	75M	Business and personal services		
(34)		Office supplies		
(35)		Packing materials		
(36)		Activities not elsewhere classified	2161	9000
		Total	27539	113995

Source: Kanamori, Takase, and Uno, Economic Growth and Welfare---An Estimation of NNW in Japan.

(2) Water pollution: Pollution of water is caused by emission of organic substance and of hazardous materials into the environment. Waste water from paper mills is an example of the former, and mercury from chemical plants is an example of the latter.

We take BOD (biological oxygen demand) as an indicator of water pollution by organic substance. Based on the outlet of industrial water by each industrial sector and the BOD content reported in Table 4, BOD discharge per unit production can be derived. Multiplied by annual value of production, total BOD discharge can be estimated.

In the estimation, BOD removal rate needs to be taken into consideration. It is judged that the removal rate had reached 70% by 1972 after an annual rise of about 5%, according to hearing from the Environmental Agency. It must be admitted, however, that it is a difficult task to obtain reliable estimates on the removal rate because of diversified processes adopted in each industry.

Taking 1955 level as the environmental target, the excess quantity of BOD discharge is calculated for the subsequent years, and the unit cost of BOD removal (Table 5) is applied in order to obtain the total cost of pollution incurred by the society. The treatment costs for food processing, textiles, pulp and paper, and chemical industries relate to the tertiary sewerage treatment, while the costs for other sectors refer to the secondary sewerage treatment. More extensive treatment is necessitated for the aforementioned sectors due to heavy pollutant contents.

(3) Industrial wastes: The industrial wastes are classified into thirteen categories according to their physical nature and possible treatment methods.

Treatment Methods	Physical Nature
(a) Combustible:	cellulose
	solid wastes
(b) Special Treatment:	garbage of plants and animals
	waste plastics and rubber
	waste oil
	feces and urine of domestic animals
	waste acid and alkali
	polluted mud
(c) Recyclable:	scrap metals
	glass
	sludge

Table 4. Industrial water and BOD in Base Year (1970)

Sector Number	Modified JSIC Code		Industrial Water Outlet (ton/day)	BOD (ppm)	Unit Treatment Cost (million yen/ton per year)
(1)	01&	Agriculture, forestry, and fisheries			
(2)	10&	Mining			
(3)	18	Food processing	4693.6	650	55
(4)	20&	Textiles	2917.5	480	69
(5)	22	Lumber and wood products	15.7	65	5446
(6)	23	Furnitures and fixtures	3.3	30	11800
(7)	24	Pulp, paper, and paper products	3278.4	750	35
(8)	25	Printing and publishing	2.6	15	14160
(9)	29	Leather and leather products	14.5	1700	76
(10)	28	Rubber products	45.9	10	1773
(11)	26	Chemicals	3158.5	150	110
(12)	27	Petroleum and coal products	178.5	30	433
(13)	30	Nonmetallic mineral products	90.5	20	2365
(14)	31A	Iron and steel materials	} 673.8	} 28	} 321
(15)	31B	Basic iron and steel products			
(16)	32	Basic nonferrous metal products	68.7	23	496
(17)	33	Fabricated metals	32.2	30	3543
(18)	34	General machinery	39.2	30	3250
(19)	35	Electrical machinery	110.1	100	713
(20)	36M	Transport equipment, motor vehicles	} 35.5	} 30	} 1187
(21)	36S	Transport equipment, others			
(22)	37	Precision instrument	4.2	30	3543
(23)	38&	Other manufacturing	16.0	15	3547
(24)	15	Construction			
(25)	70	Electric power, gas, and water			
(26)	40	Wholesale and retail trade			
(27)	50&	Finance, insurance, and real estate dealing			
(28)	50R	Real estate rent			
(29)	60T	Transportation			
(30)	60C	Communication			
(31)	97	Public administration			
(32)	75P	Public services			
(33)	75M	Business and personal services			
(34)		Office supplies			
(35)		Packing materials			
(36)		Activities not elsewhere classified			
		Total	15370.7		—

Source: Same as Table 3.

Table 5. Changes in Energy Consumption and Contributing Factors

	Total change in energy consumption	Changes due to:				
		Unit consumption coefficient	Input coefficient	Final demand	Final consumption of energy	Cross effect
Value (billion yen)						
1970 – 1975	3540.65	(–) 269.37	(–)642.77	3419.87	1335.07	(–) 302.20
1975 – 1980	1006.86	(–)1897.28	68.15	3411.18	5.01	(–) 580.15
1970 – 1980	4547.51	(–)1857.09	(–)496.68	7494.86	1340.08	(–)1933.69
Ratio						
1970 – 1975	1.0000	(–) 0.0760	(–)0.1815	0.9658	0.3770	(–) 0.0853
1975 – 1980	1.0000	(–) 1.8843	0.0676	3.3879	0.0049	(–) 0.5761
1970 – 1980	1.0000	(–) 0.4083	(–)0.1092	1.6481	0.2946	(–) 0.4252

Note: For the methodology and data, see text.

(d) Disposable as it is: cinders
rubbles

The estimation of the industrial wastes is made based on the unit waste discharge coefficients for the 13 items above in each sector and the value of annual production. The discharge of industrial wastes into environment is estimated considering the ratio of wastes escaping proper treatment. The ratio is assumed to be as high as 89% in 1965 and 74% in 1970 according to an estimation by Tokyo Municipal Office. The ratio is believed to have dropped subsequently to 20% in the case of manufacturing sectors and 50% in others.[8] Finally, the imputation of the damages to environment is done based on the cost of proper treatment for the 13 items and the quantity of improperly discharged wastes.

According to the estimated loss from environmental polution in Table 2, the increasing environmental deterioration in the 1960s have been reversed in the 1970s. It is also important to note the structural change in the economy which has taken place in the 1970s. We turn to this problem in the next section.

Conservation of Energy - An Analysis in the Input-Output Framework

In the early days of industrialization, smoke stacks were the symbols of prosperity. This remained so until quite recently, and economic growth and energy consumption were almost synonimous. We have seen that throughout the 1960s to the early 1970s economic expansion accompanied increase in energy consumption, and the elasticity was more than unity.

The relationship between growth and energy consumption marks a clear departure from the earlier trend in the last 10 years, as is shown in Figure 1. This change was undoubtedly one of the contributing factors in arresting the environmental pollution. We have witnessed conservation efforts among industries after the consecutive oil crises in the 1970s, and industrial production is now being carried out with less energy input per unit of output.

At the same time, demand has been shifting away from manufacturing toward services. As I have examined elsewhere, slower economic growth in recent years resulted in a decline in the share of private investment in plant and equipment and a concomitant gain in private consumption, the latter of which has much larger service content than the former.[9] What is more, private consumption itself is increasingly directed toward services rather than goods. The intermediate demand of industries also contain

Figure 1. Economic Growth and Energy Consumption

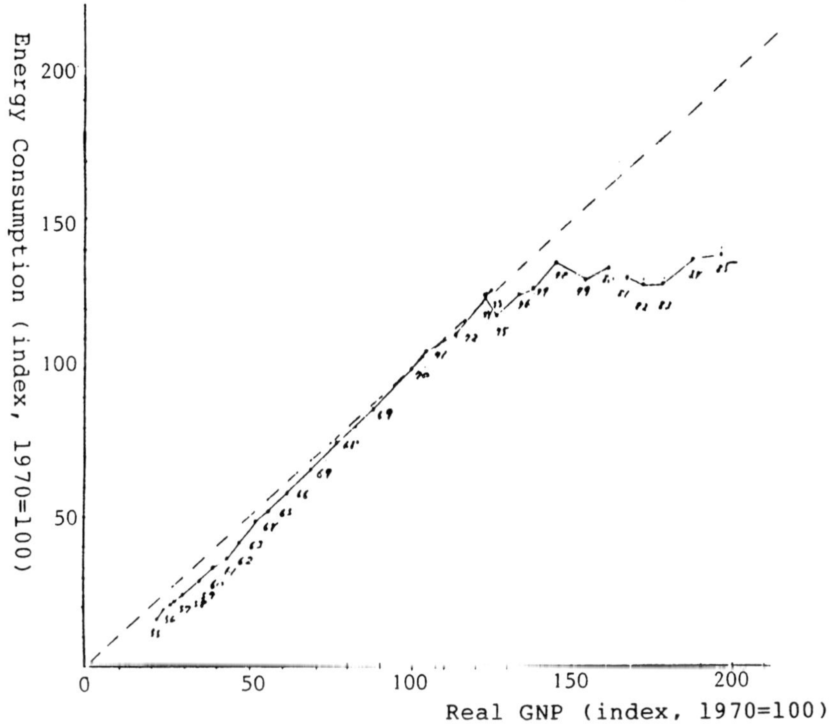

Source: GNP data are from Economic Planning Agency, Annual Report on National Accounts. Energy consumption data are from the Institute of Energy Economics, Energy Matrix.

various business services such as advertising, information services, rental services, among others. And service activities generally consume less energy per unit output.

In this section, we shall look into the contributing factors toward energy conservation. For this purpose, we employ the input-output framework which is suited to grasp the interdependent sectoral relationship among industries and final demand.[10)]

The notation used here are as follows.

B: a Leontief inverse (i*i);

H: consumption of energy per unit of production (1*1);

Q: total energy demand (1*1);

Qp: demand of energy sector, intermediate (1*i);

Qf: demand of energy sector, final (1*1);

X: value of production (1*i).

Total demand for energy is then expressed as

$$Q = \sum_i Qp + Qf. \quad \ldots\ldots\ldots\ldots\ldots\ldots(1)$$

Unit consumption of energy is definitionally derived as

$$H = Qp \;/\; X. \quad \ldots\ldots\ldots\ldots\ldots\ldots(2)$$

Thus, total intermediate energy demand can be expressed as below, where $\hat{H}$ indicates diagonal of H.

$$Qp = \hat{H}\ X. \quad \ldots\ldots\ldots\ldots\ldots\ldots(3)$$

Substituting (3) into (1) above, we obtain

$$Q = \sum_i \hat{H}\ X + Qf, \text{ or}$$

$$Q = \sum_i H\ B\ F + Qf \quad \ldots\ldots\ldots\ldots\ldots\ldots(4)$$

Therefore, denoting the base year by suffix 0 and the comparison year by t, the model for analyzing the contributing factors for changes in energy consumption is expressed as below.

$$\begin{aligned} \Delta Q &= Qt - Q0 \\ &= (\Sigma Ht\ Bt\ Ft + Qft) - (\Sigma H0\ B0\ F0 + Qf0) \\ &= \Sigma\ (H0+\Delta H)(B0+\Delta B)(F0+\Delta F) \\ &\quad + (Qf0 + \Delta Qf) \\ &\quad - \Sigma\ H0\ B0\ F0 \\ &\quad - Qf0 \\ &= \Sigma \Delta H\ B0\ F0 + \Sigma H0\ \Delta B\ F0 + \Sigma H0\ B0\ \Delta F + \Delta Qf \end{aligned}$$

$$+ \Sigma(\Delta H\Delta BF0 + H0\Delta b\Delta F + \Delta HB0\Delta F + \Delta H\Delta B\Delta F) \quad .....(5)$$

Thus, we see that total change in energy consumption is decomposed into the following five parts:

(1) due to changes in unit consumption coefficients

(2) due to changes in input coefficients which describe the structure of the intermediate inputs in an industry

(3) due to changes in the levels of final demand

(4) due to changes in final energy consumption, and

(5) due to cross product term.

The most important period to which we should focus our attention from the point of energy conservation effort is the period from 1970 to 1980 because much change had taken place during this decade. Employing the 1980 constant prices input-output tables for 1970, 1975, and 1980, which are standardized to a 36-sector classification to be consistent with a series of analysis on the Japanese economy being attempted by the present author, the changing patterns of energy consumption are examined. The results are shown in Table 5.[11)]

According to the calculation, increase in Japan's energy consumption in the 1970-1975 period amounted to 3541 billion yen (1980 prices). This reflects 3420 billion yen increase attributable to growth in final consumption of energy. The 1973 oil price hike prompted energy conservation, but the energy conservation attributable to this factor remained relatively small and is calculated to be 269 billion yen. The changing input structure was another factor which worked toward energy conservation. About 634 billion yen in the first half of the 1970s is attributable to this factor.

In the 1975-1980 period, we have witnessed a totally different picture. Although economic growth resulted in additional energy requirement comparable in value to the amount in the preceding five-year period, final consumption of energy was contained almost at the same level. Remarkable change was observed in the energy requirement per unit output, which contributed to conserve energy by as much as 1897 billion yen. Thus, of the 3411 billion yen increase in energy consumption in the last half of the 1970s, more that two thirds was offset by conservation effort, which consisted of lower unit consumption coefficient and the cross product term, and actual increase was recorded at 1007 billion yen.

If we look at the picture for the decade of 1970-1980 as a whole, the increase in final demand would have meant an additional energy requirement worth about 7495 billion yen. Increased demand in energy sector added 1340 billion yen worth energy consumption. Despite an additional energy requirement worth about 8835 billion yen, actual increase is recorded at 4547 billion yen, which is about 50% less. Energy conservation of this magnitude was made possible through reduction of energy consumption per unit production and shifting input structure (or, technological change throughout the economy), plus the term attributable to cross effect.

Concluding Remarks

In the case of air pollution, energy-saving technology contributed to lower energy demand. It has been proven by the actual working of the economic mechanism that economic growth and preservation of environmental quality can be made compatible. Before concluding the chapter, we should look at the development in pollution prevention facilities and the pollution levels.

First, as for air pollution, in addition to the energy saving technology and structural change which we have analyzed in the previous section, we have witnessed steady increase in pollutant removal capacity, as we see below.[12)]

	SOX removal capacity (million Nm3/h)	NOX removal capacity (million Nm3/h)
1970	5.4	---
1975	79.5	4.3
1980	122.0	39.1
1985	148.3	109.8

As for sulphur removal from exhaust gas, construction started in 1967 and the capacity in 1970 was recorded at 5.4 milion Nm3 per hour. By 1975, it had increased to 79.5 million Nm3/h. The capacity has doubled in the next ten years to reach 148 million Nm3/h in 1985. Coupled with other measures such as shifting to low-sulphur crude oil and sulphur removal from heavy oil, air pollution level attributable to SOX has been lowered considerably, as we shall see below.[13)] The trends for other pollutants are also listed.

	Sulphur dioxide (ppm)	Nitrogen monoxide (ppm)	Nitrogen dioxide (ppm)	Carbon monoxide (ppm)
1965	0.057			
1970	0.043	'71 0.036	0.022	'71 6.0
1975	0.021	0.028	0.026	4.7
1980	0.016	0.023	0.027	2.9
1985	0.011	0.022	0.024	2.4

Thus we see that increased provision of pollution prevention facility clearly contributed to reduction in pollution levels. For example, sulphur dioxide content, which used to be more than 0.055 ppm in the mid 1960s, has been reduced to 0.011 ppm. Similar trend is observed for nitrogen monoxide and carbon monoxide, although the situation for nitrogen dioxide leaves much to be desired.

Second, as for water pollution, diffusion of sewerage system was apparently the contributing factor in reducing the pollution level. In addition to provision of public sewerage, we should also note recycling of industrial water.[14] In order to arrest the expanding need for industrial water, Japanese industry opted to use the water repeatedly after proper treatment, contributing to the reduction in pollution levels.

	Area under sewerage system (hectares)	Industrial water (1000m3/day) (% recycled)
1962		37,320 (27.8)
1965	75,760	49,162 (36.3)
1970	131,998	85,042 (51.7)
1975	214,069	121,625 (67.0)
1980	308,547	138,850 (73.6)
1985	447,422	n.a.

	BOD levels in urban rivers (ppm)	Water containing hazardous materials (%)
1970	'71 16.6	1.4
1975	10.5	0.17
1980	8.6	0.05
1985	6.4	0.02

The two series of data, one referring to biological oxygen demand representing general pollution of water by organic substance and the other referring to percentage of water samples containing designated materials which are harmful to human health, are consistent in showing improving trends in water pollution.[15] The data are available only after the situation became alarming around 1970, but the trends thereafter show considerable improvement.

Third, as for disposal of waste, one can cite statistics on total output and planned disposal as follows.[16)]

	Total output of waste (ton/day)	Planned disposal (ton/day) (percent incinerated)
1968	62,005	54,271 (48.3)
1970	76,998	69,899 (55.3)
1975	87,167	76,273 (57.8)
1980	120,371	113,728 (60.4)
1985	117,916	112,590 (69.1)

The first series which shows the total output of waste concerns the amount of waste output in the area covered by collection plans by local public bodies. In 1968, this covered about 75% of the population. By 1975, the coverage was expanded to virtually 100%. The second series concerns the amount collected by local public bodies, and the difference between the two series of data was disposed of individually, often improperly. We see that this gap is being narrowed recently. At the same time, the portion of waste being incinerated, which is a preferred way of waste disposal, shows steady increase. For example, the ratio in 1968 is recorded at about 43%, compared to 69% in recent years. In earlier years, a larger percentage of waste went into land fill or sea reclamation.

Regarding industrial waste it is important to note that an increasingly larger portion is being recycled. According to a survey by the Ministry of International Trade and Industry, 58.5% of total was recycled in 1983.[17)]

Reflecting general improvement in environmental pollution, grievances against environmental pollution has actually declined in recent years. Figure 2 shows the trend. It is observed that the total number of grievances increased rapidly in the 1960s through early 1970s. The fact that the rate of increase exceeds economic growth rate implies the serious nature of the problem. In 1972, the number of grievances exceeded 15,000 cases, including those for air, water and soil pollution, noise, vibration, and bad smell. In the 1980s, the number is recorded at around 9,000, or a decline of about 40% compared to the peak years.

However, one cannot be overly optimistic. For example, complaints against noise and vibration have not shown similar improvement, indicating the difficulty in coping with this pollution category based on the efforts in the private industry. In view of the fact that much of these cases are attributable to less than adequate zoning in Japan, a different kind of efforts is called for, and the public sector has the greatest responsibility. Increase in 'other' causes for complaint suggests emergence

Figure 2

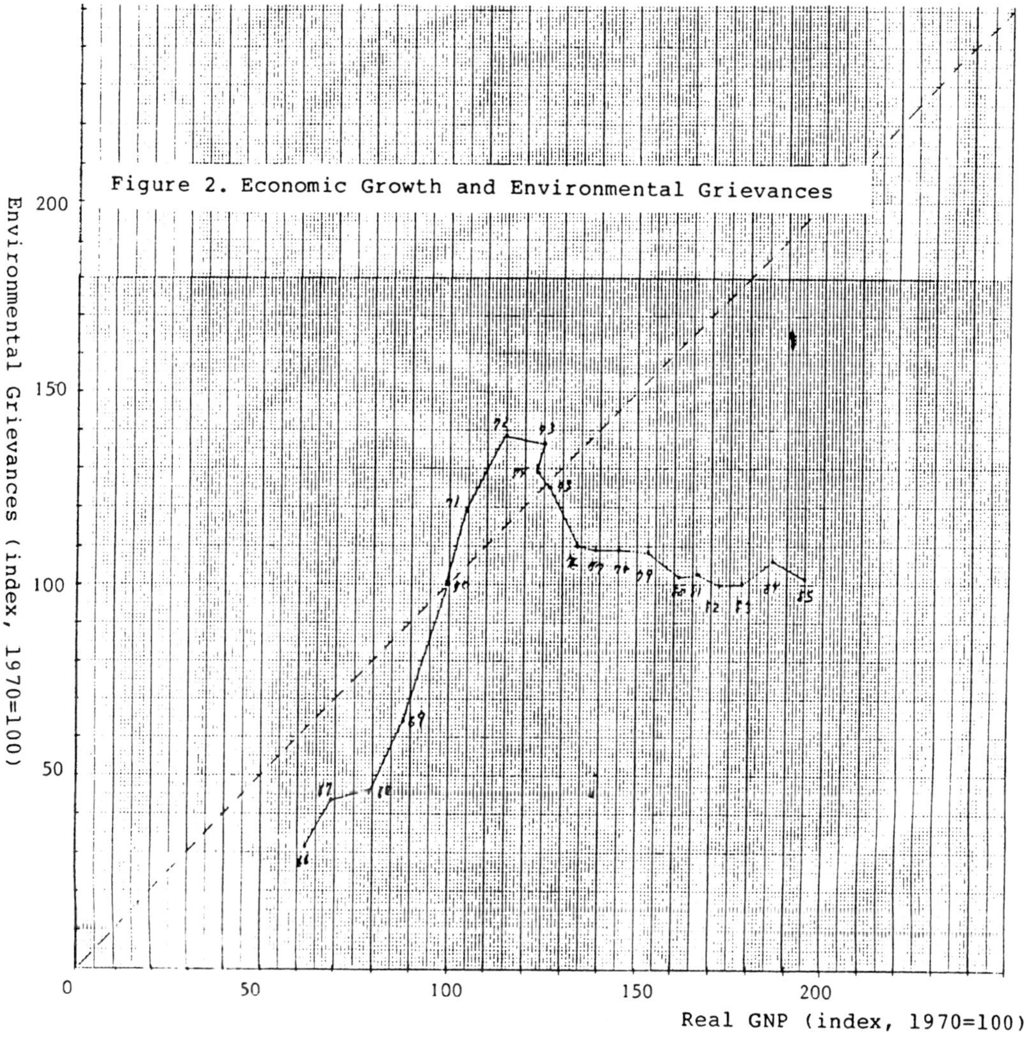

Figure 2. Economic Growth and Environmental Grievances

Source: Environmental Disputes Coordination Commission, Survey on Grievances against Environmental Pollution.

of new types of pollution.

Thus, according to the Japanese experience, economic growth and upkeeping of environmental quality can be made compatible. There are two factors working to this end; pollution prevention and structural change of the economy toward high-tech and services.

Finally, we should note that policy is formulated only after the need is recognized, a factor which economists call the 'recognition lag'. Unlike purely economic variables such as growth rate and unemployment which are regularly monitored, this lag can be extremely long. In the case of Japan, it was perhaps as long as ten years, judging from the individual cases of environmental conflicts and the aggregate observation such as trends in NNW. And the policy was shifted only after bitter struggle by those who were adversely affected by the pollution. We should keep watchful eyes on opinion of the people in view of the emergence of new types of environmental disputes. Improper land use in urban districts and careless destruction of natrual environment by public investment projects are the cases in question.

Notes

(1) Kanamori (1977).

(2) In updating the NNW estimates some changes were made as to data sources, which include the following. (a) The government consumption is now based on SNA classification by social purpose entitled 'government final consumption expenditure according to cost-composition and purpose'. The purpose of such changes was to make the calculation of NNW consistent with the framework of revised SNA. (b) The household purchase of durable goods is also based on SNA. (c) The estimation of environmental damages is based on the actual levels of pollution. In earlier estimates, this item was derived from the volume of input of hazardous materials into economic activity, implicit assumption being that they approximate the emission of such materials in the absence of removal process. Due to improvement in production process in subsequent years, however, we have observed higher removal rates but total effect has not been estimated. In cases where unit coefficients were employed, such figures often have come to overstating the actual figures in recent years due to technological change, or became irrelevant due to changes in production processes. Thus, it seems more realistic to estimate the environmental damages based on the actual pollution levels relative to the base year (1970); the cost to attain improvement can be estimated, based on the investment data for pollution prevention facilities. (d) When the new estimates deviate from the earlier ones in absolute levels, newer figures were linked to the base year (1970) figure. In general, however, the discrepancy is minimal.

(3) Kanamori (1977).

(4) Bureau of Statistics, Basic Survey on Social Life, various issues.

(5) The Japan Broadcasting Corporation (1986), p. 1000. For detailed analysis of working hours by industry, see Uno (1987), Chapter 8, pp. 147-158.

(6) Uno (1987), Chapter 16, pp. 271-310.

(7) Uno, ibid., p. 302.

(8) Economic Council (1973), p. 151.

(9) Kimio Uno, 'Measurement of Service Orientation in Production and Employment in an Input-Output Framework'. Paper presented at the 20th General Conference on the International Association for Research in Income and Wealth, Rocca di Papa, 1987. See also Kimio Uno, 1989.

(10) The methodoloy here follows Administrative Management Agency, 1980 Input-Output Tables, Explanatory Report, p. 213.

(11) The data are from Management and Coordination Agency (1985).

(12) The data are from Environment White Paper, various issues.

(13) The figures are averages of observation points for which data are available througout the period under consideration. The data are from Environment White Paper, various issues.

(14) Sewerage data are from the Public Facility Survey by the Ministry of Home Affairs. The data for industrial water are from the Census of Manufactures by the Ministry of International Trade and Industry. The industrial water listed here refers to plain water only, and does not include sea water.

(15) BOD levels are average of 8 urban rivers. As for hazardous materials in water, the figures refer to the percentage of water samples containing designated hazardous materials.

(16) Total output refers to the amount of waste produced within the planned collection area. The data are from Environment White Paper and Handbook on Environment Statistics, various issues.

(17) Environment White Paper, 1987 edition, p. 258.

References

An asterisk (*) indicates that the title is the translation by the present writer. Otherwise, the English title is from the original publication.

Bureau of Statistics, various issues, Basic Survey on Social Life (Shakai Seikatsu Kihon Chosa).

Economic Council (1973), New Welfare Indices - NNW* (Atarashii Fukushi Shihyo - NNW), Ministry of Finance Printing Bureau.

Environment Agency, various issues, Environment White Paper* (Kankyo Hakusho), Ministry of Finance Printing Bureau.

Environment Agency, various issues, Handbook on Environmental Statistics* (Kankyo Tokei Yoran), Gyosei.

Environmental Disputes Coordination Commission, various issues, Report on the Survey on the Number of Grievances against Environmental Pollution.

The Institue of Energy Economics, various issues, Energy Matrix, (Enerugi Matorikkusu).

The Japan Broadcasting Corporation (Nihon Hoso Kyokai), (1985*), Time Budget of the Japanese People, Illustrated, (Zusetsu Nihonjinno Seikatsu Jikan, 1985). Nihon Hoso Shuppan Kyokai, 1986.

The Japan Society of Industrial Machinery Manufacturers, various issues, Production Statistics of Pollution Prevention Equipment* (Kogai Boshi sochino Seisan Jissekini Tsuite).

Kanamori, H. (1977), Japanese Economic Growth and Economic Welfare Center Pepar No. 31, The Japan Economic Research Center.

Kanamori, H., Takase, Y, and Uno, K. (1977), Economic Growth and Welfare - An Estimation of NNW in Japan* (Keizai Seichoto Fukushi - Nihonno NNWno Suikei), Research Report No. 41, the Japan Economic Research Center.

Management and Coordination Agency (1985), 1970-1975-1980 Link Input-Output Tables (Showa 45-50-55nen Setsuzoku Sangyo Renkanhyo), National Federation of Statistical Associations.

Mine, N. (1976), The Estimation of Some Basic Data about Industrial Pollution, (J) The Journal of Japan Economic Research, no. 5, Mar.

The Ministry of International Trade and Industry, various issues, Trends of the Industrial Pollution Prevention Investment in the Private Sector* (Minkanni Okeru Sangyo Kogai Boshi Toshino Suiii).

Nordhaus, W.D, and Tobin, J. (1972), Is Growth Obsolete? National Bureau of Economic Research, General Series 86, Economic Growth, New York and London: Columbia University Press.

North, D.C. and Miller, R.L. (1971 and 1973), The Economics of Public Issues. New York: Harper and Row.

Organisation for Economic Cooperation and Development (1974), Subjective Elements of Well-Being. Paris.

Organisation for Economic Cooperation and Development (1977), Environmental Policies in Japan. Paris.

Uno, K. (1987), Japanese Industrial Performance. Amsterdam: North-Holland.

Uno, K. (1989), Measurement of Services in an Input-Output Framework, Amsterdam, North Holland.

Chapter 18

COMPARATIVE ANALYSIS OF ENVIRONMENTAL POLICIES

'Comparisons are odorous'
Shakespeare, Much Ado About Nothing

G. Scimemi[1]

Introduction

The entire range of activities carried out by international organisations is heavily dependent on comparative policy analysis. This is the case for the environment as much as, or perhaps even more than, for other policy fields. And it is the case for the OECD not less than for other international agencies. In order to illustrate the point, one may wish to look, sequentially, at the set of components which, taken together, form the OECD Environment Committee mandate, and give some thought to what role is played by comparative analysis, relative to each of these tasks. The current mandate of the OECD Environment Committee includes the following seven items:

- examining **common problems** related to environmental management and proposing effective means of preventing, minimising or solving them, taking into account economic and other relevant factors;
- encouraging the **harmonization** of environmental policies among the Member countries;
- proposing policy options to prevent or minimise **conflicts** that could arise between member countries as the result of **national environmental policies**;
- proposing policy options to prevent or minimise **conflicts** that could arise between Member countries in the use of **shared environmental resources**;
- reviewing, assessing and consulting on **environmental measures taken or proposed** by Member countries;
- **assessing trends** in environmental quality and improving the information base for environmental and natural resource management policies;

[1] The opinions expressed and the arguments employed in this paper are those of the author and do not necessarily represent those of the OECD.

- giving special attention to ways and means by which Member countries can contribute to environmental improvement and better management of natural resources in countries **outside the OECD region.**

Seven Tasks for Comparative Environmental Policy Analysis

A first function of international co-operation is the identification of **common problems** and the search for **common solutions**. Indeed, when comparative analysis is systematically carried out within a family of sufficiently similar countries, this tends to trigger imitation and emulation efforts, and significant opportunities for learning from example may be the result. A case in point is the flourishing international diffusion of environmental impact assessment, a particularly successful policy instrument, first developed in the United States and subsequently adopted in an increasing number of countries. This diffusion process also illustrates certain difficulties, which the transfer of experience may meet, because of differences in national institutional cultures. One can especially perceive a tendency to rely on traditional instruments such as land-use planning and zoning regulations, where these (contrary to the U.S. case) have worked in a satisfactory manner in the past.

Another function of international co-operation has to do with the **harmonization** of environmental policies. Harmonization is an essential prerequisite for international trade, especially between countries linked by important import/export flows of commodities subject to environmental regulations. An important case in point is chemicals. Trade in chemicals would be altogether impossible if countries involved were inadequately aware of the respective policies: the role of comparative policy analysis is evidently essential. What is more, if each country were to impose its own standards, its own regulations, its own laboratory practices and its own testing methods on each new chemical product entering the market, this would impose unbearable costs, delays and uncertainties on every producer, exporter, importer or user. The OECD Chemicals Programme, initiated ten years ago, copes with this problem with considerable success, drawing on all Member countries' experience to select best control options and methods, harmonizing regulation, establishing mutually accepted packages of data, mutually accepted laboratory practices and whatever else is needed in this domain.

But harmonization does not necessarily mean uniformity; there are cases where choosing different policy approaches in different countries may be fully justifiable. Thus a third important function of international cooperation is to ascertain that differences in policies, or policy measures, are based on acceptable reasons. When and where this is not the case, there is a legitimate suspicion that some governments may use strict environmental regulations in order to protect their industry and create non-tariff barriers against imports: this may result in **conflicts between Member countries**. One example is the regulation of car emissions. Comparative analysis shows immediately that countries, and especially those possessing an important automobile industry, while selecting the modalities and strictness of emission control measures, are often influenced by their present or prospective markets, both domestic and international. The risk of seeing emission regulations used as a tool for a protectionist strategy is the original reason for the long-standing involvement of the Commission of the European Communities in the matter. From the standpoint of eliminating barriers to trade, the task has been to avoid excessively strict, not excessively lenient environmental criteria. Only after the recent adoption of the so-called Single Act has the Communities' mandate been expanded beyond free trade matters and now includes environmental concerns as such, and as well. This recent development may now allow for certain countries to motivate and obtain permission to tighten their standards beyond the levels established by the Commission.

International co-operation is also required to minimise **transboundary pollution**, a different potential source of **conflict**. Again, comparative analysis of existing and proposed policies performs an essential role in the development of the necessary international arrangements. In the case of the transboundary transport of hazardous waste, a subject on which the OECD is currently working, in order to define an internationally agreed list of those materials which constitute hazardous waste, and must hence undergo control, the OECD has had to analyse the current approaches in a dozen countries, plus the EC Directives, plus several international conventions on sea pollution. Indeed, experience shows that to try and settle transfrontier pollution problems while failing to take into account national policies, past and present, and national conditions, both in the emitting and in the receiving countries, invariably leads to further problems. This has been the case for the so-called Helsinki Protocol which has established a single round figure of 30 per cent for reductions of sulphur oxide emissions and transboundary flows, to be achieved by all signatory countries irrespective of their previous efforts, and regardless of the pollution loads they were imposing upon their neighbours. Given this obvious lack of rigour, both from the equity and the efficiency viewpoint, certain countries (including important countries) have easily found good reasons for abstaining. The mistake has been noted, and it has been recommended that it should not be repeated concerning the Nitrogen Oxides Protocol, presently under development at the Economic Commission for Europe of the United Nations.

Whatever may be the nature of the problems at hand, and I have mentioned just a few, countries eventually settle them through the collective acceptance of principles, agreements or international directives. But then there is a need to **monitor the policy measures** whereby individual countries implement them. That is one additional essential role of comparative analysis. OECD regularly monitors the implementation of the polluter-pays principle in Member countries, as required by the Council. Considerable differences between countries are frequently found in the interpretation both of the rule and of the admissible exceptions. The Commission of the European Communities has a special need -and an explicit mandate- to monitor the implementation of the Communities directives. Failure to conform to a directive may occur both by defect (as in the case of the exceedingly high levels of atrazine in water tolerated in Italy) and by excess (as in the case of the excessively rigorous requirements imposed by Greece on imported milk powder). Comparative evaluation of implementation always requires very careful interpretation. Thus, for example, a recent enquiry done for the Commission excluded Germany, France and The Netherlands from the list of countries which were abiding to the Seveso directive. What this actually means is that those countries had actually anticipated the action taken in Brussels and had enacted their domestic regulations in the area of industrial accidents, prior to the action taken by the Commission. What they had failed to do, then, was merely to modify those regulations to bring them into conformity with the procedure subsequently established in Brussels.

Comparative analysis is not only used to identify differences. It may also help to detect certain similarities between countries, with respect to their approaches to environmental policies. In the case of OECD Member countries, substantial similarities emerge, largely attributable to the relative homogeneity of conditions prevailing in western industrialised nations; one can even distinguish three successive phases in what is, broadly speaking, a **common trend** of policy development in this set of countries. During a first phase, industrialised countries have set up legislation, institutional and regulatory frameworks, essentially to deal with pollution problems. During this phase, which broadly coincides with the

1970s, these countries have enacted laws, adopted principles, such as the polluter-pays principle, and established environmental ministries and agencies. A second phase followed, during which they focussed on the implementation issues under consideration to include natural resource management, risk management, and the identification of new pollutants. This phase covers the 1980s. A third phase has recently commenced, in which the emphasis is on the integration of environemtnal concerns in a number of important policies in various sectors, such as energy, transport, agriculture, industry and international trade. At the same time, the range of concerns further expands to include ecological and economic interdependence between developed and developing nations.

This **opening outside the OECD region**, specifically to address Third World problems is, no doubt, the most striking feature of the current 'third' phase. It is frequently attributed to the recent publication of the Report of the World Commission on Environment and Development. One must be aware, of course, that precedents for this new orientation had been already established in a number of international bodies, notably the OECD Environment Committee - see its special session of April 1st, 1981, and the two Reports 'Challenges fo the Eighties' and 'Economic and Ecological Interdependence' issued in 1981 and 1982 respectively. This new dimension is likely to gain considerably from exercises of comparative policy analysis. It is unfortunate that, up until today, efforts in this direction have been mostly inadequate, both in quantity and quality. To the extent that they exist, comparative studies have, more often than not, simply looked at conventional methods, previously tested over a small number of countries, and tried to extend them on the planetary scale. It has been often forgotten that such an extension entails a radical and original reconsideration of comparative analysis, its nature, its scope as well as its limitations. The degree of homogeneity, which is the lot of the small group of 'rich' countries, certainly does not apply on the global scale. This has substantial implications, methodological as well as normative. When adopting a global perspective (as opposed to a regional viewpoint, such as the one prevailing in the previous sections of this discussion) nearly every point which has been made in the foregoing paragraphs concerning the objectives and the applications of comparative analysis needs to be fundamentally questioned again. To fail to recognise this need means to risk falling victim to a number of optical illusions, each such illusion being, in some way, the distorted image of one or other of the previously discussed objectives.

From a Regional to a Global Perspective

To begin with the concept of 'common problems' and 'common solutions', one is immediately confronted with the reality of wide-ranging diversities between countries within the Third World. How many common features could one hope to depend upon? Geographic, cultural, economic and institutional conditions tend to be extremely dissimilar. The conventional thesis of the 'development stages' which assumes the existence of a similar development pattern shared by all developing countries except for longer or shorter time lags, has generally lost its credibility. Thus the scope for immediate transfers of experience, for direct learning form mutual example, even among countries comprised in the same broad and loose category, is more the exception than the rule. Worse still is the indiscriminate attempt to transfer certain approaches from the 'North' to the 'South', and certain economic principles from advanced market economies to countries where the market's role is marginal and/or severely distorted.

As concerns harmonization, one should be aware of the need to avoid the fallacy of world-wide uniform standards and norms. In a variety of local geographic and anthropological conditions which is several times wider than what is known in the so-called First World, this fallacy, based on misguided humanitarian concern, and a misunderstood concept of equal rights, risks encroaching upon each government's prerogative, each country's entitlement to judge what is good for its people, including deciding upon the appropriate trade-off between environment and other socio-economic imperatives, both long and short-term.

Moving then to the issue of non-tariff barriers, one should not forget, of course, that the concern expressed by Lesser Developed Countries at Stockholm, to the effect that environmental regulations in rich countries may become an obstacle to import from the 'South' is still there. But it would be misleading to ascribe primarily to these regulations the origin of existing major trade barriers. Much more serious market distortions stem from straightforward protectionist measures, which have little to do with environmental requirements and which severely penalise Third World exporting countries.

As concerns conservation of shared resources, what needs to be avoided is the fallacy of one-dimensional solidarity. Based on the premise that resource conservation is in the common interest of the North and the South, the supporters of this fallacy tend to ignore that, depending on the country under consideration, the balance between the weight of sacrifices and the package of benefits may greatly vary. Thus, for example, solidarity will not thrive along the dimension of natural resources conservation, unless equity also grows along the dimension of welfare. In the absence of adequate compensatory flows, whoever talks in terms of 'our common future', or 'only one earth' sould at least be aware that rich and poor are bound to receive his or her message in very different ways.

The comparative assessment of implementation measures is another risky exercise, when one tries to cover vast and remote countries, for several of which information concerning such measures is only indirectly available through the channel of understaffed and underequipped governmental offices, episodical expert reports and formal policy declarations (which often reflect wishful thinking and good intentions, but not necessarily the reality).

Similarly, the monitoring of the state of the environment and its trend is a particularly fragile exercise in countries where statistical data, even on basic national indicators, are scarcely reliable, and where vast extents of territory are not even mapped, let alone surveyed and monitored for environmental and resource conditions.

The conclusive illusion is that of the so-called 'universal goals of humanity': a notion which invariably tends to embrace concepts essentially generated within the contemporary western culture, and which become automatically endowed with the virtues of universality and rationality, thus ignoring one of the very basic lessons of cultural anthropology. 'We now have the critical mass', said Dr. Tolba, Executive Director of UNEP last week in Nairobi, 'a consensus over what sustainable development implies'. Indeed, such consensus was probably shared by the distinguished members of the UNEP Governing Council he was addressing. But what about reaching beyond that rather atypical audience? The concept of sustainable development, as well as its practical implications, are far from being collectively shared, in the majority of countries, and by the majority of their population, at the level of grass roots, where the action is. Even in developed countries, fewer people than one might hope understand the notion of global sustainable development and accept what this entails in terms of radical changes in production and consumption patterns, attitudes, lifestyles and societal goal systems.

All these warnings are not intended to discourage extending comparative environmental policy analysis beyond the boundaries of the limited world regions we know fairly well, but to point out the magnitude of the challenge and the risk of arriving hastily at entirely misleading conclusions.

Some Final Remarks

It is tempting, of course, to try and apply comparative analysis with a view to ranking countries with respect to their standard of performance as concerns environment and environmental policies. Such temptation, however, should rather be resisted; results can only be controversial at best, detrimental at worst; there are both technical and diplomatic reasons for that.

For one thing, every attempt at ordinal ranking postulates the existence of a single indicator. But environment is essentially a composite good, including components as diverse as air quality improvement, accident prevention, landscape enhancement, and preservation of genetic diversity. The boundaries of the concept also vary to such an extent that elements such as seismic risks or safety in the workplace are included by some, excluded by others. Even if appropriate indicators were to be defined for each factor included, one could hardly conceive of reducing the resulting multi-dimensional vector to a single significant parameter.

This may be just one of several methodological problems. On top of these there are also matters of fair play and good taste (or shall I say savoir faire?). A distinction needs to be made -at least on grounds of diplomacy- between a judgement which a government brings concerning its own successes and failures, and the assessment that other countries or some international body might feel authorised to make. There is nothing wrong with the Nordic countries, for example, taking pride in their successes; one cannot but appreciate the amount of effort they put into taking upon themselves the mission of ecological pioneers and models: and the value of the demonstration effect they achieve. Other countries -allow me not to mention which ones- do not hesitate to admit that they lag behind everybody else; this may help to raise public awareness and stir some sense of responsibility for improving things domestically, especially if the media pick up the point. On the other hand, the idea for an international collegial jury or for an international secretariat to distribute good grades and bad grades must be discouraged. It has been wisely noted, e.g. that it would not be advisable for representatives of European governments to spend their time at international meetings 'pointing national fingers at each other in some sort of ugliness contest to find the dirtiest man in Europe'. This statement is remarkable in more than one respect. First, the message it conveys. Secondly, the occasion. It was Prince Charles who pronounced it, a few weeks ago, opening the Ministerial Conference on the North Sea. 'Comparisons are odorous': in the country of Shakespeare one does not fail to savour the subtly ironic scent of this curious lapse which escapes from a character in one of his most celebrated plays.

Selected Bibliography

American Bar Association & The Institute for European Environmental Policy (1987), (Conference on) Environmental Regulation in the European Community and the United States, Brussels, February.

Brickman, R., Jasanoff, S. and Ilgen, T. (1982), Chemical Regulation and Cancer: A Cross-National Study of Policy and Politics, New York.

Enloe, C. (1975), The Politics of Pollution in a Comparative Perspective: Ecology and Power in Four Nations, New York.

Greco, N. (1987), Processi decisionali e tutela preventiva dell'ambiente, Economica Pubblica, 17, 12.

Gusman, S., Von Moltke, K., Irwin, F. and Whitehead, C. (1980), Public Policy for Chemicals: National and International Issues, Washington D.C.

Haigh, N., Bennet, G., Kromarek, P. and Lavoux, T. (1986), European Community Policy in Practice, vol. 1, Comparative Report; Water and Waste in Four Countries, London, Graham and Trotman.

Hourcade, J.C. (1988), Enjeux économiques et technologiques des politiques de l'environnement en Europe: contribution à une évaluation comparative des politiques publiques de l'environnement au sein de la CEE, CIRED.

Kelman, S. (1981), Regulating America, Regulating Sweden: A Comparative Study of Occupational Safety and Health Policy, Cambridge.

Knoepfel, P. (1981), Comparative Analysis of the Implementation of SO_2 Air Quality Control Policies in Europe, Conceptual Framework and First Results, Berlin.

Knoepfel, P. and Weidner, H. (1983), Implementing Air Quality Control Programs in Europe: Some Results of a Comparative Study, Boston.

Knoepfel, P., Weidner, H. and Hanf, K. (1979), The Implementation of SO_2 Control Policy, in: Chemical Policy in the European Community and Switzerland, Aspen.

Knoepfel, P., Weidner, H. and Hanf, K. (1980), International Comparative Analysis of Program Formation and Implementation in SO_2 Air Pollution Control Policies in the EEC Countries and Switzerland: Analytical Framework and Research Guidelines for the National Research Teams, Berlin.

Kunreuther, H. and Linnerooth, J. (1982), Risk Analysis and Decision Processes: The Siting of LEG Facilities in Four Countries, Laxenburg.

Lundqvist, L. (1980), The Hare and the Tortoise: Clean Air Policies in the United States and Sweden, Michigan.

McLoughlin, J. and Forster, M.J. (1982), The Law and Practice Relating to Pollution Control in the Member States of the European Communities: A Comparative Study, Graham and Trotman.

Mangun, W.R. (1977), The Public Administration of Environmental Policy: A Comparative Analysis of the United States and West Germany, Bloomington.

OECD (1980), Siting Procedures for Major Energy Facilities: Some National Cases, Paris.

OECD (1981), The Environment - Challenge for the '80s, Paris.

OECD (1982), Economic and Ecological Interdependence, Paris.

OECD (1988), The OECD Chemicals Programme, Paris.

Rehbinder, E. and Stewart, S., Environmental Protection Policy, in: Cappelletti, Seccombe and Weiler, (eds.), Integration Through Law, Europe and the American Federal Experience, De Gruyter, Berlin, New York.

Vogel, D. (1980), Coercion Versus Consultation: A Comparison of Environmental Protection Policy in the United States and Great Britain, Denver.

Whetstone, G. and Rosencranz, A. (1983), Acid Rain in Europe and North America, Washington D.C.

INDEX